The Scottish and Alaskan Offshore Oil and Gas Experience and the Canadian Beaufort Sea

AF678664

The Scottish and Alaskan Offshore Oil and Gas Experience and the Canadian Beaufort Sea

J.G. Nelson
Sabine Jessen

Canadian Arctic Resources Committee, Ottawa, Canada
Faculty of Environmental Studies,
University of Waterloo
Waterloo, Ontario, Canada

© 1981, Canadian Arctic Resources Committee
and Faculty of Environmental Studies,
University of Waterloo
Quotation with appropriate credit is encouraged.

Nelson, J.G. (James Gordon), 1932-

The Scottish and Alaskan offshore oil and gas
experience and the Canadian Beaufort Sea

Co-published by University of Waterloo, Faculty of
Environmental Studies.
Bibliography: p.
Includes index.
ISBN 0-919996-22-1

1. Offshore oil industry — Scotland — Shetland Islands.
2. Offshore oil industry — Alaska — North Slope.
3. Offshore oil industry — Beaufort Sea region.
4. Energy development — Scotland — Shetland Islands.
5. Energy development — Alaska — North Slope.
6. Energy development — Beaufort Sea region.
I. Jessen, Sabine. II. Canadian Arctic Resources
Committee. III. University of Waterloo. Faculty
of Environmental Studies. IV. Title.

HD9560.N44 333.8'231 C81-090110-2

Initial funding for this report was provided by the
Department of the Environment and the Department of
Indian Affairs and Northern Development.

Canadian Arctic Resources Committee
1981 Publishing Programme
J. Craig McNaughton, Publications Director

CARC's 1977-1982 Publishing Programme
is made possible by a grant from the
Richard and Jean Ivey Fund.

Photos: Courtesy of J.G. Nelson, Dan Kowalski-Lightwriting, Inc.,
Ian K. MacNeil and Parks Canada.

Contents

List of tables

List of figures

Acronyms

BLM	Bureau of Land Management
CARC	Canadian Arctic Resources Committee
CEQ	Council on Environmental Quality
COPE	Committee for Original Peoples' Entitlement
CZM	Coastal Zone Management
DCRA	Department of Community and Regional Affairs
DFG	Department of Fish and Game
DIAND	Department of Indian Affairs and Northern Development
DNR	Department of Natural Resources
DOC	Department of Commerce
DOI	Department of the Interior
DPDP	Division of Policy Development and Planning
EEC	European Economic Community
EIA	Environmental Impact Assessment
EIS	Environmental Impact Statement
EPA	Environmental Protection Agency
FY	Fiscal Year
GAO	General Accounting Office
GB	Great Britain
NOAA	National Oceanic and Atmospheric Administration
OCS	Outer Continental Shelf
OCSEAP	Outer Continental Shelf Environmental Assessment Program
OCSESP	Outer Continental Shelf Environmental Studies Program
SDD	Scottish Development Department
SESP	Socio-economic Studies Program
SOTEAG	Shetland Oil Terminal Environmental Assessment Group
SSSI	Sites of Special Scientific Interest
SVEAG	Sullom Voe Environmental Assessment Group
USGS	United States Geological Service

Drilling rig off Alaskan shoreline, Dan Kowalski-Lightwriting, Inc., Juneau, Alaska

Summary

Beaufort Sea offshore oil and gas exploration and development have been underway for about five years and are at a critical stage. Proposals are currently being considered for very large-scale development — including twelve or more offshore gas fields, and construction of artificial islands, production platforms, pipelines, coastal roads, buildings, and harbour facilities. The potential impact of such development on the environment, renewable resources, and existing land uses is great. Such impact is of special concern in the Beaufort because the area is of national and international importance for wildlife and other renewable resources — notably beluga whales and snow geese. The Beaufort coastal area is also of considerable importance to Inuit and other traditional trappers, hunters, and long-time inhabitants of the region.

In seeking guidance on how to manage potential Beaufort Sea petroleum development, it is useful to study the Scottish and Alaskan experiences. Both Scotland and Alaska have been subjected to oil and gas development pressures for about a decade, with many deep-sea wells and an array of port and other coastal facilities having been constructed along the Scottish coast, mainly in the last five years. On the other hand, little offshore construction has occurred in Alaska because of careful strategic planning, environmental impact assessment, and other management procedures. In Scotland, the Shetland Islands are of particular interest because of their isolation, their physical and socio-economic characteristics, and because of the development there of a successful management approach.

An analytical framework is necessary to compare and derive lessons from management experience in different areas. The management assessment model used in this report consists of four parts: agency, planning, implementation, and general guides or characteristics. Each of these parts can be broken down in a hierarchical way as a basis for describing and analyzing a management system (see Table 1). Thus planning can be said to include activities such as strategic planning, inventory, classification, and evaluation. Implementation includes approval procedures, construction, and operation; these in turn include surveillance, inspection, monitoring, enforcement, and other processes. The general guides or characteristics are applicable to all major aspects of management and include such processes as research, co-ordination, and information access.

Through chapters two to five this management assessment model is applied to the Alaskan and Scottish management experiences, with particular emphasis on the North Slope Borough and the Shetlands. Information for this part of the report was obtained through a literature review and interviews, particularly in Washington, D.C., Anchorage, Juneau, and various places in Scotland.

Table 1
Management assessment model

	Government or Management Level or Type		Local	Provincial	Federal	International	Committee	Commission	New Agency	Industry	Other
Agency	Lead Agency										
	Participating Agency										
Planning	Strategic Planning										
	Inventory (Biophysical & Social)										
	Land Classification										
	Especially Significant Areas										
	Evaluation										
Implementation	Approval	Negotiated Development									
		Pre-Hearing									
		Hearing									
		Formal Case Record									
		Permit Order With Conditions									
		Appeals									
		Final Decision									
	Construction and Operation	Surveillance & Inspection									
		Monitoring									
		Enforcement									
		Modification									
General Guides or Characteristics	Research										
	Co-ordination	Government									
		Industry									
		Public									
		Inter-group									
	Economic Incentives										
	Information Access										
	Management Process Monitoring										
Comments											

Nelson (1980)

Many similarities and differences can be observed in offshore petroleum development and its management in the Alaskan and Scottish cases. Development has been associated with many legal and institutional changes in both areas — for example, in planning for activity on the continental shelf. A strong local response to offshore development has occurred both in the Shetlands and in the North Slope Borough. In the Shetlands, this response has occurred largely within the context of a modified pre-existing local government system. In the Alaskan case, greater stress has been placed on relatively new comprehensive senior government legislation, such as the Coastal Zone Management Act of 1972. However, a central purpose of this act has been to provide the financial and technical assistance needed to stimulate more comprehensive and effective state and local management.

A concern for more co-ordinated and effective sea-use planning (Young and Fricke, 1975) has developed in both countries, largely as a result of recent offshore petroleum development pressures. In the marine and coastal context, the United States has made considerable progress in anticipatory or strategic planning, as well as with inventories of shipping, dumping, fishing, and other activities. Resources have been classified and mapped, and especially significant areas have been identified. Environmental impact assessment and other evaluation procedures are used more in the U.S. than in Scotland. There is a strong interest in improved monitoring in both areas, in large part because of a common acceptance of the many technical, economic, and environmental uncertainties facing managers. Inspection, surveillance, and other procedures during construction and operation are inadequate in both countries.

Chapters six and seven focus on a more detailed study of the Shetlands. The discussion is comparative and is undertaken in the context of three basic management strategies: (1) insulation and containment; (2) integration and absorption; and (3) major social change. The first two strategies have been emphasized in the Shetlands. Similarities among the Shetlands, the North Slope Borough, and the Canadian Beaufort region include: isolation; low population; little urbanization; large dependence on fish and other renewable resources; human occupancy for thousands of years; distinctive regional ethnic characteristics in the population; a homeland view among the residents; a wilderness or wildland view among outsiders; and a succession of whaling, mining, and other invasions of the "boom and bust" type, culminating in the recent large-scale quest for petroleum.

Differences include: (1) large-scale land ownership by the Shetlanders and Alaskans, but not by the Inuit and other Beaufort Sea residents; (2) a strong local government system in the Shetlands and the North Slope Borough, but not in the Canadian Beaufort area; and (3) greater financial and technical resources, and regional self-reliance among the Shetlanders and Alaskans than among the people of the Canadian Beaufort region. Senior and local government is fragmented in all three areas with greater progress on this problem being made in the Shetlands and the North Slope Borough than in the Canadian Beaufort Sea area.

Information in chapter seven focuses on the effectiveness of the

Shetlands oil development and environmental management system. Forecasting errors have generally led to underestimations of the extent of the population, the labour force, housing, and other forms of oil-induced growth. Road development, pipeline, and other impacts are being managed reasonably well. Fisheries are under heavy pressure, not only because of the loss of access to fishing grounds, damage to equipment, and the inflationary effects on the labour force, but also because of the synergistic effects of over-fishing and other changes attributable to pressure largely from the European Economic Community. Eight oil spills have occurred at the Sullom Voe oil terminal since operations began in late 1979, one of which was major and caused many bird deaths and other losses. Problems have been experienced with the discharge of tanker ballast and other transportation effects.

An effective coastal zone management system has been developed in the Shetlands. The Sullom Voe oil terminal harbour authority, which reports to the Shetland Islands Council, has responsibility for the entire Shetlands coast to the three-mile limit. The authority has an eighty-man staff, a £20 million wharf and support complex, and a large operations budget financed from pilot charges, oil and gas production fees, and other payments provided for under the Zetland County Council Act 1974, which gave the Shetland Islands Council special powers to manage oil and gas development effects. A regional planning system has also been developed to prepare and update official plans, district plans, project approvals, and other details — again largely on the basis of oil and gas income.

Chapter eight contains a summary and recommendations for the Canadian Beaufort Sea. The recommendations are built on the conclusion that land ownership, strong local self-government, and associated financial, technical, and other benefits have supported a self-reliant, intelligent, and effective response to oil development in the Shetlands and Alaska and can also do so in the western Canadian Arctic.

This conclusion leads to the recommendation that certain basic principles should be implemented as soon as possible in the Beaufort Sea-Mackenzie Delta area:

1. Strengthen and improve local government, preferably by establishing a regional system of some type for Inuvik, Tuktoyaktuk, and other centres;
2. Improve the permit or project approval system, preferably by introducing a more comprehensive approval method with a stronger role at the regional level;
3. Allocate land to the native people and to the territorial governments as a basis for entrepreneurial and investment activities, and stewardship;
4. Establish a Beaufort Sea-Mackenzie Delta shore zone planning commission: to co-ordinate the various agencies; to conduct strategic planning; to arrange for inventories, and land and water classifications; to identify sensitive or environmentally significant areas; to monitor and to conduct research. This commission should have a specified life-span and yield eventually to a strengthening regional government. The commission could, however, retain certain research, technical,

monitoring, and other overall support functions for a longer period;

5. Create an Arctic Environmental Forum, with representatives from all levels of government, industry, and the public, including environmental groups such as the Canadian Arctic Resources Committee (CARC) and the Association of Canadian Universities for Northern Studies (ACUNS);
6. Improve Canada's shore zone co-ordination system by adding a clearing-house function and staff and other requirements needed to provide better information and research for Canada's new coastal and marine frontier.

Acknowledgements

This report was prepared originally for the National Co-ordinator of the Shore Zone Integrated Program, Environmental Management Service, Department of the Environment. We would like to thank the members of the Study Steering Committee — Michael Parkes, James Maxwell, Ron Bailey, and Bill Brakel — for their guidance and support. As co-ordinator of the project, Michael Parkes was especially helpful. Gratitude is also extended to the people who assisted with the research, many of whom are listed in Appendix 3. Special thanks goes to Murray Walsh, Professor George Dunnet, and Ian Townsend Gault for reviewing the report.

The results were presented by J. G. Nelson in March and April 1981 to the Study Steering Committee, to a group of senior DOE officials, to the Task Force on Beaufort Sea Developments, and to the Senior Policy Committee, Northern Development Projects. We thank the members of these groups for their questions and comments. Appreciation also goes to Bonnie Kent, Chris Matulewicz, Bonnie Lee Bender, and Marko Dumancic for ensuring successful production of the report. Financial support from the Department of the Environment (DOE) and the Department of Indian Affairs and Northern Development (DIAND) made the research possible. Don Gamble of the Canadian Arctic Resources Committee (CARC) was helpful in launching the study and has been supportive throughout. The views expressed in the report are not necessarily those of the departments of the Environment and Indian Affairs and Northern Development.

Finally, we would like to express gratitude to Craig McNaughton of CARC, who edited this volume. His attention to detail and his energy and tenacity have done much to improve the quality of this publication. We are of course responsible for its imperfections.

Mackenzie Delta, Parks Canada

Chapter I: Introduction

Beaufort Sea development

Beaufort Sea development is currently at a critical stage. The most promising oil deposits have been found by Dome Petroleum, although Gulf Canada Resources and Esso Resources Canada have had some success in the area. Dome has recently put forward an overall development concept that includes twelve or more offshore oil fields, construction of artificial islands, production platforms, pipelines and oil collection systems, and a wide array of coastal roads, buildings, erosion control structures, and dredging sites. Dock, warehouse, residential, and other facilities have already been built in Tuktoyaktuk. Dome is now considering construction of other ports and services on Parry Peninsula and Cape Bathurst, as well as at King Point. Various companies have proposed gas processing plants at localities in the Beaufort Sea-Mackenzie Delta area.

Impacts

Bottom deposits and aquatic and coastal environments could be disturbed almost continuously over perhaps a forty-year oil and gas production period. Whale and fish populations could be placed in jeopardy, including the threatened bowhead and beluga whales which are valued highly by the Inuit. Bird populations could be placed at risk. Of particular concern are oil spills during staging periods when much of the continental bird population of species such as the snow goose is resting in the Mackenzie Delta and north Yukon coastal areas. Construction and development could also have a negative effect on grizzly bears, polar bears, seals, and other aesthetically and economically important animals. Adverse effects on such animals would reduce the food supply of the Inuit, as well as hunting and tourist opportunities for residents of other parts of North America.

The Beaufort Sea and the nearby Mackenzie Delta are unique in the Canadian Arctic, as are wildlife and other resources associated with these two areas. It is most important, therefore, that every effort be made to plan Beaufort Sea oil and gas development so as to minimize adverse impacts on these resources and on the human activities dependent upon them. Possibilities include some sort of Beaufort Sea development authority, a regional government, or a coastal zone management system of the type used in attempts to bring industry and governments together in other parts of the world.

Figure 1 **Beaufort Sea area**

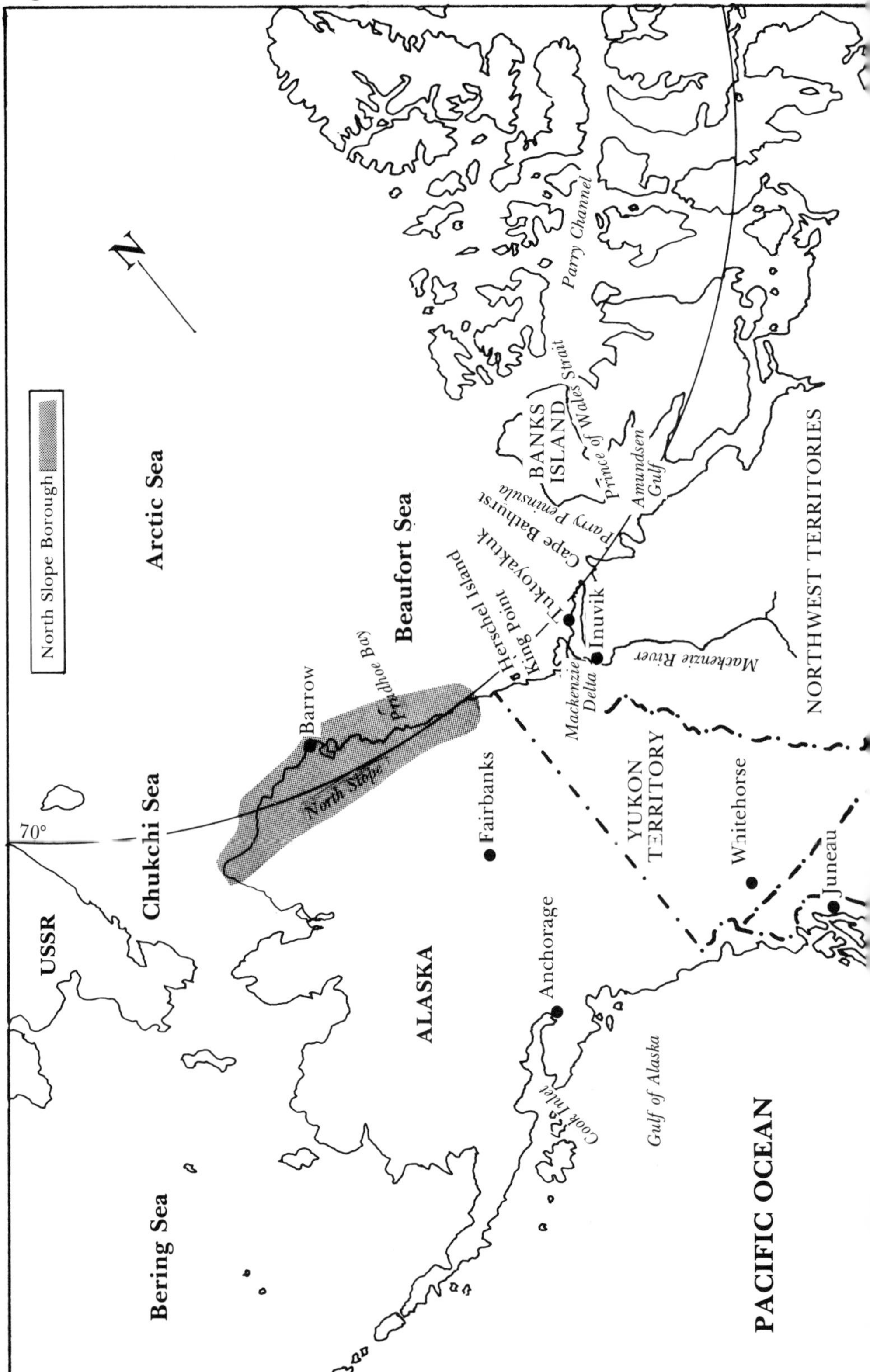

The Scottish experience

In seeking guidance on how to approach the management of potential Beaufort Sea petroleum development, it is useful to study the Scottish experience. As a result of large-scale North Sea oil and gas discoveries during the last decade, Scotland has undergone pressures and changes similar to those now underway in the Arctic and in other parts of offshore Canada. Unique management arrangements have been developed in the Shetland Islands, providing for oil and gas production while at the same time attempting to control unwanted impacts on fishing, tourism, and other pre-existing industries and the renewable resources and environment upon which they are dependent. Other arrangements have been made in the Orkney Islands and at Inverness and other communities along the Scottish east coast. The advantages and disadvantages of these arrangements can provide lessons for Canada.

The Alaskan experience

We can also profit from the experience of the Americans in Alaska. The North Slope Borough is of special interest in that industrial developments there could have a very direct effect upon the Canadian section of the Beaufort Sea. An outstanding example of this potential is the recent proposal by Globtik Tankers (USA), Inc. for an integrated arctic marine transportation system to ship oil and gas to eastern and southern markets in Canada and the United States. To move two million barrels a day from a Prudhoe Bay or alternate North Slope terminal through the Beaufort Sea, Amundsen Gulf, Prince of Wales Strait, and the Northwest Passage reportedly would involve twenty-four icebreaking tankers of 350,000 tons each, coupled with thirty-one smaller 80,000-ton tankers.

A second reason for being interested in the Alaskan situation concerns the attempt there to introduce an integrated coastal zone management system under the Coastal Zone Management Act of 1972. This legislation provides a planning and cost-sharing framework for comprehensive coastal zone management, including provision for energy facilities as well as protection of renewable resources and the environment.

Analytical framework

A management model is required as a basis for comparing other experiences with the Beaufort Sea situation. Very few models of comprehensive management systems have been developed. The model used in this report has been inspired by renewable resource studies for Inuit Tapirisat of Canada (Nelson, 1975); environmental impact assessments of the Alaska Highway pipeline (Alaska Highway Pipeline Panel, 1977); and a recent evaluation of the environmental protection regulatory system for the Nanticoke industrial complex in southern Ontario (Nelson, Day, and Jessen, 1980).

Figure 2 Scottish North Sea area

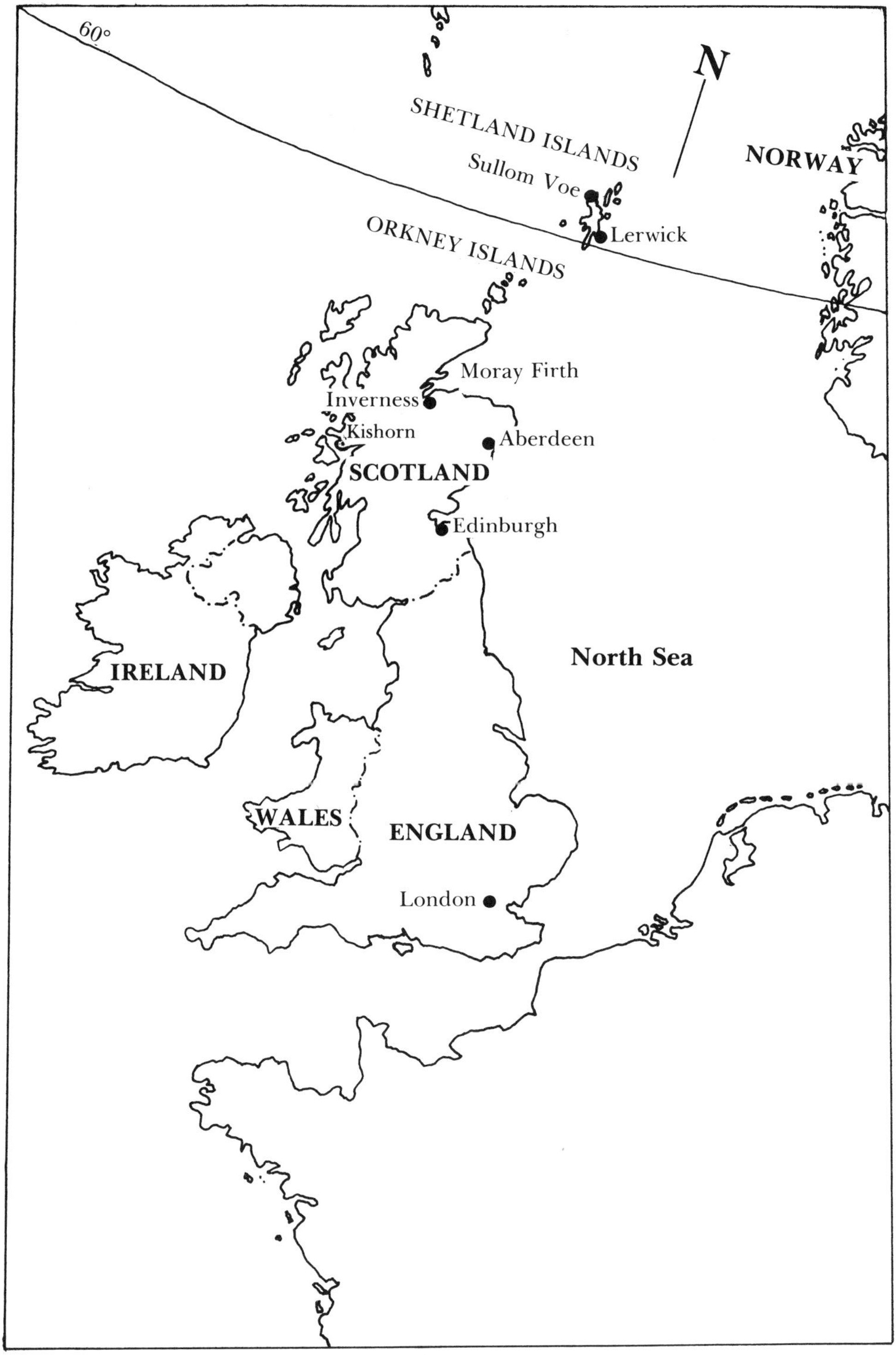

The elements of the model (Table 1) are not intended to indicate order or sequence in planning and management. The steps that are taken and the feedback loops involved will vary from situation to situation. The model is also likely to be amended in various ways as it is applied to different situations. Nevertheless, it provides a useful framework for the description and the analysis of the management efforts of the major interested parties or actors — including those at different levels of government, and those in business, native, and other groups. The questionnaire in Appendix 1 explains more fully the various parts of the model.

Research plan and schedule

Canadian information has been obtained mainly through a review of literature, and interviews in corporate, government, university, and other offices in Ottawa, Calgary, Edmonton, and Whitehorse. A questionnaire (Appendix 1) was distributed to members of a number of federal departments, and to other organizations and individuals concerned with the Beaufort Sea. Relatively few responses were received. A critical view of the current Beaufort Sea management system was submitted by Don Gamble of the Canadian Arctic Resources Committee in response to the questionnaire, and is included in Appendix 2 because of its comprehensiveness and its perception of the issues. An array of useful articles and commentaries by government, university, and informed persons can be found in the two volumes of *Northern Transitions* (Peterson and Wright, 1978; Keith and Wright, 1978). An evaluation of environmental regulation of oil and gas projects in the Arctic has recently been completed by Hunt and Lucas (1980). It indicates that many *ad hoc* committees and departments are involved in an overlapping and apparently rather inefficient regulatory and management system.

Information on the Scottish experience was obtained by the first author while on sabbatical leave in Europe in late 1980 (see Appendix 3), and during a previous study tour three years ago. Interviews and field studies were also conducted in the Shetland Islands in September 1980. Considerable information on the Scottish experience and on petroleum development and its effects on the European North Sea coast was secured at an offshore oil development conference in Edinburgh in late September 1980. Interviews and library work in Washington, D.C. in August 1980 also provided data on oil and gas development, its management, and its effects in various parts of the world, including the United States. A short September 1980 visit to Anchorage and Juneau by the second author as part of other research work also provided some information on the Alaskan experience with petroleum development in the Gulf of Alaska and on the Alaskan North Slope.

Resource, time, and other limitations have prevented an understanding of the Alaskan North Slope situation comparable to our understanding of Scotland and the Shetlands. In particular, it has not been possible to secure information on the actual effects of oil development on the environment, resources, and land uses in the North Slope Borough and in other parts of Alaska, nor on the way in which these effects have been controlled or mitigated by the management systems in use there. Such information is available for the Shetlands as a result of field work in these islands.

Report organization

This report is organized around the Management Assessment Model (Table 1). Chapters two, three, four, and five are devoted to the main divisions of the model: agency, planning, implementation, and general guides or characteristics. Most of the information pertains to Scotland, although much is also included on Alaska. In chapter six, the experience in the Shetlands — and to a lesser extent, Alaska — is considered in greater detail and some major points of comparison are made about the environment and the economic, social, and political circumstances within which oil development has taken place. Chapter seven includes detailed information on the major oil development effects in the Shetlands and the extent to which these have been controlled by the management system. The final chapter contains a summary of the major principles that have emerged, especially from the Shetlands experience. Recommendations are also made for the management of Canadian Beaufort Sea oil and gas development.

Water quality control facilities and oil storage tanks at Sullom Voe oil terminal, September 1980, J.G. Nelson

Chapter II: Agencies

Introduction

Both local and senior governments in Scotland and Alaska work closely with industry in managing oil development. A number of public groups have also become involved in various oil development proposals and the effects of proposed developments, although their legal mandate is less clear than is that of either government or industry. Various committees or other bodies have been set up to link government, industry, and public groups (discussed more fully under the section on co-ordination in chapter five). The present chapter is intended to provide a succinct outline of the evolving role of the major governmental and other actors; new laws and other changes arising from the pressures of oil development, notably in the coastal zone; and also of major new agencies such as the Shetland Oil Terminal Environmental Advisory Group (SOTEAG).

Scotland

In practice, senior government in Scotland functions as a fragmented two-part system (Batey and Breheny, 1978; G.B., Central Office of Information, 1979; Houston, 1978; Kellas, 1980). This situation results from the creation of the Scottish Office in Edinburgh, which houses five key senior government departments for Scotland. Certain of these departments — for example, the Scottish Economic Planning Department (SEPD) and the Scottish Development Department (SDD) — have no exact counterparts in the rest of the United Kingdom. The Scottish Office is headed by the Secretary of State for Scotland, who is a member of the U.K. parliament and Cabinet. He is assisted by two ministers of state. Further details on senior government organizations are shown in Table 2. A recent referendum on a separate legislative assembly and other powers for Scotland resulted in a negative decision. It is not clear, therefore, how senior government will evolve in the future.

Local government

Local or municipal governments in Scotland generally consist of elected regional and district councils. These councils are basically responsible for land use planning and implementation — for housing, and for industrial, educational, recreational, and other uses of their areas. Responsibility for air

Table 2
Central administration in Scotland, 1978; principal functions of government agencies

UK

Treasury
Inland Revenue
Customs and Excise
- Taxes, duties, interest rates
- Public expenditure

Department of Trade
Department of Industry
Department of Employment
Department of the Environment
Ministry of Transport
- Railways, ships

Ministry of Defence
Department of Energy
- Oil, coal, gas

Department of Health and Social Security
- Social security only

Department of Prices and Consumer Protection
Department of Education and Science
- Universities and research councils only

Ministry of Agriculture, Fisheries, and Food
- Animal health, food only

Home Office
- Immigration, cruelty to animals

Foreign and Commonwealth Office
- Passports

Civil Service Department
- Recruitment and conditions of civil service

Secretary of State for Scotland

Responsible for the Scottish Office, consisting of the following departments:

Scottish Home and Health Department
- Law and Order
- National Health Service

Scottish Education Department
- Education (except universities and research councils)
- Social work
- Libraries
- The Arts

Department of Agriculture and Fisheries for Scotland
- Land settlement, crofting
- Forestry
- Agricultural price supports (UK/EEC)
- Agricultural education
- Fisheries

Scottish Development Department
- Local government
- Housing
- Roads
- Transport

Scottish Economic Planning Department
- Economic planning
- Industrial incentives
- Manpower services
- North Sea oil support services
- New towns
- Electricity
- Highland development
- Tourism

Central Services
- Finance and personnel

Lord Advocate

Lord Advocate's Department
- Legal advice to government

Crown Office
- Public prosecutions

Source: Kellas, 1980

and water quality and certain other environmental concerns is shared with the Scottish Development Department, other departments of the Scottish Office, and central government departments such as the Department of the Environment. Powers of regional and district councils chiefly derive from the Town and Country Planning (Scotland) Act 1972. Other pertinent legislation includes the Local Government (Scotland) Act 1973, which provides for regional planning reports, and the Community Lands Act 1975 which, subject to authorization by the Secretary of State for Scotland, empowers a local authority to acquire by agreement any land which in its opinion is suitable for development.

The present system of local government in Scotland is a relatively recent innovation. The British government had decided by the mid-1960s that reforms were urgently required in the structure of local authorities in England and Wales, Scotland, and, a little later, Northern Ireland. Royal commissions were set up, with the commission for Scotland headed by a senior Scottish judge.

The resultant report, submitted in 1969, pointed to the extraordinary confusion of jurisdictions exercised by different bodies, and to the disparities in size and function between the many and various local authorities — all the result of decades of tinkering with, rather than reforming, the system. It recommended a completely new two-tier structure, a recommendation which was in accord with the conclusions of the commission for England and Wales, and which was eventually implemented by the Local Government (Scotland) Act 1973.

The government made some changes when implementing the report, but the end product was essentially as recommended. The mainland of Scotland was divided into nine regions, corresponding to historical, cultural, or similar divisions; each was subdivided into a number of districts of roughly equal area or population. The importance of preserving local historical associations was stressed in the report and implemented where possible; the old county of Fife, for example, which had been a kingdom centuries before, now became a region, albeit with incursions from surrounding regions. A large town or city now became a district.

This system applies to the mainland only. It was clear that with respect to the Western Isles, the Orkneys, and the Shetlands, that this system would be untenable due to the smaller areas and populations involved. Assimilation of island groups by onshore regions would have gone against the philosophy of retaining historical associations wherever possible; the new system was designed to bring people closer to their local government. This could hardly be done, in the case of the Shetlands, by assimilating the group into the Highland region, with the principal seat of administration hundreds of miles away in Inverness. Separate islands councils were established for the three areas, which bodies fulfil in one tier the responsibilities met by two tiers onshore.

According to the Town and Country Planning (Scotland) Act 1972, the basic planning objective of the various planning authorities is to determine the best use of land in the interests of the community. The means of doing this are through the preparation of regional and district plans and also by requiring developers to submit project applications for planning permission. Structure (policy or strategy) plans are prepared by the region (county in England). Local plans basically reflect regional policies through zoning and are prepared by district councils. Planning permission for development projects is granted by district councils on a case-by-case basis in the context of district and regional plans. A proposal can be approved even if a site is not appropriately zoned, although generally with difficulty. Such an approval requires council direction, endorsed by the Secretary of State for Scotland, or council amendment of the district plan through a successful appeal to the secretary.

Local governments generally have no special role in the shore zone, although unusual powers can be acquired for harbours and their management — for example, through the Private Legislation Procedure (Scotland) Act 1936, as has been done by the Shetland Islands Council. The locus of power and responsibility for petroleum development in the coastal zone has shifted in the last decade, at first towards a stronger role for senior government (the Department of Energy and the Scottish Office). More recently, the trend has been to place greater stress on local approval through district councils.

Senior government

At the senior government level, a complex division of responsibilities characterizes the presently evolving system. Key agencies in the Scottish Office are the Scottish Economic Planning Department and the Scottish Development Department. The significant role of these agencies in housing, transport, and other matters is apparent in the data found in Table 2. The division of responsibility between these departments and also between them and local government is not always clear and has been complicated by petroleum development.

As a result of a mid-1970s dispute about planning permission jurisdiction for nearshore oil platform construction facilities at Portavadie, Loch Fyne, the Court of Session ruled that municipal planning authorities had no powers below the low-water mark (Houston, 1977: 74-75). Various senior government departments are therefore now basically responsible for both shallow- and deep-water development in complex association with local government bodies and the often-powerful harbour authorities. Each statutory harbour authority derives its power from its own private legislation. While the various acts frequently incorporate provisions similar to those in the parent Harbours, Docks, and Piers Clauses Act 1847, they also include different powers for shore and other types of management. These differing harbour authority arrangements present opportunities to co-ordinate the diverse senior and local government powers in more effective ways — a point to be discussed later in the context of the Shetlands experience.

Some senior government shore and marine responsibilities are exercized under older legislation, as well as under a number of more recent statutes passed largely because of the North Sea oil and gas development pressures of the last ten years (Houston, 1977; G.B., Scottish Information Office,Reference Unit, 1974; G.B., Central Office of Information, 1979). Older jurisdictions which apply in the nearshore zone include: those of the crown estate commissioners, who are responsible for protection of recreation, navigation, fishing, and other forms of public access to the shore and the bed of inland waters (Crown Estate Act 1961); and those of the Department of Trade for any works installed below the high-water mark, inasmuch as these works could affect erosion, deposition, and shore formation (Coast Protection Act 1949).

More recent acts and regulations of significance in regard to offshore oil and gas development and its effects on the shore zone include: (1) the Continental Shelf Act 1964; (2) the Prevention of Oil Pollution Act 1971; (3) the Control of Pollution Act 1974; (4) the Dumping at Sea Act 1974; (5) the Petroleum and Submarine Pipe-lines Act 1975; and (6) the Offshore Petroleum Development (Scotland) Act 1975. On the basis of such legislation, various offshore petroleum development and environmental management responsibilities have been assigned rather imperfectly to an array of departments in London and Edinburgh. These arrangements have changed through time and are currently in flux (G.B., Scottish Office, SDD, 1977c).

Under the Continental Shelf Act 1964, the rights to explore in and exploit petroleum in the U.K. offshore area are vested in the Crown and are controlled from London. The Department of Energy is responsible for licensing policy and for granting licences to explore and to produce. For production licensing purposes, the shelf has been divided into blocks measuring 200-250 square kilometres (80 - 100 square miles). Decisions on block allocation reportedly involved relatively little input from other agencies or the public — in contrast to the Alaskan approach discussed below.

The Prevention of Oil Pollution Act 1971 deals with a wide range of air, water, and other environmental considerations. The act provides for heavy penalties in the event of deliberate oil discharge into the sea from offshore operations. Its clauses and those of related acts are not considered so clear-cut and enforceable in the case of accidental discharges. Here judgements on negligence need to be made. The Secretary of State for Trade is said to be "responsible for legislation governing the control of oil and chemical pollution, primarily from ships, and for remedial action at sea" (G.B., Scottish Office, SDD, 1977e). The Secretary of State for Energy is responsible for controlling offshore petroleum development operations, including the pollution therefrom. Requirements relating to well-drilling methods, blow-out prevention, and related matters are attached to exploration and production licences. The Department of Energy's Petroleum Production Inspectorate is charged with enforcing these requirements and regulations.

The Secretary of State for Scotland is said vaguely "to have responsibilities in relation to oil and chemicals on Scottish beaches" (G.B., Scottish Office, SDD, 1977e). On the other hand, although not bound statutorily to do so, local authorities have reportedly accepted responsibility for clearing oil from beaches, banks of estuaries, and up to about one mile from shore. Action on oil spills in docks and harbours is usually co-ordinated by the concerned harbour authority (G.B., Scottish Office, SDD, 1977c).

The Minister of Transport has major responsibilities for commercial harbours and ports (G.B., Scottish Office, SDD, 1977e). Scottish harbours are dealt with from the department's Ports Office in Edinburgh. Among other things, the department ensures proper consideration of all aspects of a commercial port, including administration and control, geographical suitability of the site, and economics. All harbour works costing more than £1,000,000 must have approval of the Minister of Transport. The department can make loans to help finance approved projects, and also provides engineering and economic advice on schemes of all sizes. No mention is made of resource or environmental advice.

The loading of wastes for dumping at sea is reportedly controlled by the Department of Agriculture and Fisheries for Scotland (DAFS) through the Dumping at Sea Act 1974 (G.B., Scottish Office, SDD, 1977e). This act prohibits permanent dumping of articles and substances in the sea and tidal waters except in the case of a licence that covers activities such as construction works (piers, harbours, and coast protection facilities).

Two key recent statutes are the Petroleum and Submarine Pipe-lines Act 1975 and the Offshore Petroleum Development (Scotland) Act 1975. Under the former act, the central government in London has powers to approve the route, size, and other characteristics of submarine pipelines, as well as responsibilities for their effects on resources and the environment. Certain responsibilities are assigned to the oil companies; for example, they are required to consult fishing interests on any submarine pipeline routing before a works authorization is issued by the Department of Energy. How such consultation works in practice and how it bears on government decisions are not clear. The act also gives the Secretary of State for Energy power to authorize construction of large refineries with a capacity of more than 1,000,000 tons of crude liquid petroleum *per annum* prior to any approval by local planning authorities. However, no such stipulation applies to large gas refineries.

The Offshore Petroleum Development (Scotland) Act 1975 provides senior government and the Department of Energy in particular with an array of powers. By issuing a sea designation order for an offshore area, the department can subject all petroleum-related developments below the low-water mark in that area to regulation and licensing, thus dealing with the planning permission difficulty raised as a result of the Court of Session decision regarding Portavadie, Loch Fyne. Powers are also provided to require industry to set aside and provide funding for rehabilitation of the site.

Wide powers are granted to the Department of Energy to acquire land for any purpose (except refining) relating to offshore petroleum exploration and development. These powers were introduced because of senior government concern in the early 1970s about whether local governments would approve sufficient land to accommodate the perceived high demand for construction of oil production platforms. Of special interest were the large concrete platforms requiring very deep water — as is found along the scenic Scottish west and north coasts.

In addition to providing compulsory land acquisition powers, the Offshore Petroleum Development (Scotland) Act 1975 allows for fundamental changes in normal planning permission procedures to ensure that the national interest in petroleum development cannot be delayed or blocked by regional or local concerns. For certain major types of facilities such as platform construction, shore oil terminals, roads, and housing, compulsory acquisition can be completed in as little as twenty-one days and without the normal safeguard of a public inquiry in the case of objections by persons whose land is being taken. This expediting procedure has not been used, largely because a high demand for platform sites has not materialized. Resentment over the autocratic nature of these procedures also has been an influence in this regard.

New agency: Shetland Oil Terminal Environmental Advisory Group (SOTEAG)

One major new organization is the Shetland Oil Terminal Environmental Advisory Group (SOTEAG), which was set up in the summer of 1977 by the Sullom Voe Association (SVA) (Dunnet, 1980). The SVA is a non-profit-making, limited liability company created in May 1975 under the powers acquired by what is now called the Shetland Islands Council pursuant to the Zetland County Council Act 1974. The SVA is responsible for carrying out the construction, development, and operation of the new Sullom Voe oil terminal. Its authorized share capital is divided into shares held equally by the Shetland Islands Council and the thirty-one oil companies participating in the Ninian and Brent North Sea oil fields and the pipelines and facilities at Sullom Voe.

The membership of SOTEAG is divided into two classes: executive and associate. The executive consists of a chairman and a secretary from the University of Aberdeen, two representatives of the Shetland Islands Council, and two members from the oil companies. Nine associate members have been appointed from: (1) the National Environment Research Council; (2) the Nature Conservancy Council; (3) the Countryside Commission for Scotland; (4) the Department of Agriculture and Fisheries for Scotland; (5) the Industrial Pollution Inspectorate; (6) the Scottish Development Department; (7) the Shetlands Fisherman's Association; (8) the Shetlands Bird Club; and (9) the University of Aberdeen. There are also observers from agencies such as the Sullom Voe terminal project (industry) and the Shetland Islands Council's Planning Department. SOTEAG also has an eight-person monitoring committee (several members of which are not from SOTEAG), as well as an industry observer, and the SOTEAG secretary. The composition of SOTEAG is thus representative of major interest groups and of the professional expertise required to address its task. SOTEAG's terms of reference are:

> To examine and advise on the environmental implications of the terminal at Sullom Voe during the future stages of construction (including site rehabilitation), commissioning, and operation. The area for review shall be restricted to within the zone for which the Shetlands Council is responsible for environmental matters and will include the activities of tankers to Sullom Voe (Foxton, 1979).

The inclusion of the last statement on tankers extends the area of SOTEAG operations considerably.

The tasks of the monitoring committee are:

1. To advise SOTEAG on a monitoring strategy that will allow early detection of environmental change resulting from the oil terminal and associated developments during the construction and operational phases. Consideration of effluents and emissions to the environment and the adequacy of this monitoring will be a necessary part of this responsibility.
2. To define the requirements of monitoring programmes, to develop proposals to meet these requirements, and to make recommendations to SOTEAG on their implementation.
3. To initiate additional studies in relation to environmental damage and recovery in the event of a major accident.
4. To interpret and assess the results of these programmes and other relevant information and report to SOTEAG.
5. To be alert to any future oil-related developments in the Shetlands and provide early advice on appropriate monitoring programmes (Foxton, 1979).

Industry

An array of private and public corporations are active in the oil development field in Scotland. Certain of the public corporations such as the National Coal Board or British Gas Corporation are statutory undertakers whose operations are subject to special legislation such as the Coal Industry Acts 1946 to 1977. This legislation often gives the developer special powers — for example in land expropriation — although private companies can gain access to expropriation powers through local government or the Private Legislation Procedure (Scotland) Act 1936. In the case of British Gas Corporation, the national gas distributor in the U.K., such legislation exempts the corporation from certain planning requirements — such as securing district council planning permission for underground pipelines. However, as compressor stations and other above-ground facilities are subject to local planning permission, this exemption tends to be an imperfect one.

Voluntary and other organizations

It is impossible to describe adequately the many voluntary or non-statutory organizations involved in Scottish land use, resource, and environmental matters relating to oil development (G.B., Scottish Office, SDD, 1977e). Scotland and the U.K. as a whole have long traditions of public consultation, principally for directly affected parties — those whose lands or economic activities are to be interfered with in a fundamental way, notably by compulsory purchase. Third-party intervenors who are concerned about particular issues such as wildlife, or about policies that affect many people in wide-ranging ways, are not as well provided for in the law or by custom. According to some observers of the Scottish scene, government sees itself as an adjudicator in such matters, but this stance is less easy to accept as government agencies become more involved in road, harbour, energy, and other development activities.

Some major voluntary organizations have been formed recently, in part as a result of the effects of oil development in Scotland. An example is Friends of the Earth, originally a U.S. organization, which is becoming more and more active internationally. Another of the significant new agencies is the Scottish Forum on the Environment. The voluntary bodies on this forum include the Association for the Protection of Rural Scotland, the Conservation Society, the Council for British Archaeology, the National Farmer's Union of Scotland, the National Trust for Scotland, the Royal Society for the Protection of Birds, the Scottish Civic Trust, the Scottish Council of Social Services, the Scottish Landowners Federation, and the Scottish Wildlife Trust. Associated with these organizations on the forum are three senior government agencies with comparable interests: the Countryside Commission for Scotland; the Nature Conservancy Council; and the Scottish Tourist Board. Some of these organizations have large memberships: the National Farmer's Union of Scotland has about 20,000 members; the Scottish Wildlife Trust, a relatively new agency, has about 7000 members in twelve Scottish branches; and the Royal Society for the Protection of Birds has over 250,000 members, although they are found throughout the United Kingdom.

Alaska

Under the federal Coastal Zone Management Act of 1972, the U.S. government provides funding and technical assistance for the delimitation, planning, and management of the coastal zone. The basic responsibility for such planning and management is left to the states, with the federal government retaining the authority to approve state coastal zone management plans. In 1977, the Alaskan state legislature passed the Alaska Coastal Management Act, and in 1979 the federal Secretary of Commerce approved the Alaska Coastal Management Program (ACMP).

Local government

The Alaska Coastal Management Program stipulates that all coastal local governments, or districts which exercise planning and zoning powers, must prepare coastal management programmes. In essence, these programmes are outgrowths of traditional community planning and zoning activities, although they embrace additional considerations, such as habitat and resource protection (U.S., Department of the Interior, USGS, Conservation Division, 1980: 71).

In developing their coastal programmes, local governments are provided with financial and technical assistance from the state through the Alaska Coastal Management Act. These programmes must be approved by the state legislature. An important criterion for approval is that the

> district plan should not unreasonably or arbitrarily restrict or exclude uses of state concern, which include the use of resources and the siting of facilities for energy production in the coastal zone (U.S., Department of Interior, USGS, Conservation Division, 1980: 74).

On the other hand, federal and state energy projects and activities in the coastal zone must be consistent with the local plan once it has been agreed to by the various levels of government. Theoretically then, national, state, and local issues are worked out in the district plan stage, rather than the project stage. The effectiveness of this consistency provision as a basis for reconciling senior and local government concerns is still uncertain as few evaluations of its utility have as yet been conducted.

As is noted in the 1979 annual report on the Alaska Coastal Management Program:

> The particular method chosen by the Legislature to effectuate coastal planning and management in Alaska was to establish a partnership between the state and its local governments. The principal obligation of local governments is to develop comprehensive coastal resource management programs for the coastal areas under their jurisdiction.... For the state's part, each agency is charged with assisting local governments in formation of their programs, and with complying with the programs once they are approved (State of Alaska, Office of the Governor, DPDP, Office of Coastal Management, 1980a: 2).

Three Alaskan local or district coastal management programmes were approved in January 1980. A fourth by the North Slope Borough "was withdrawn for further refinement" (North Slope Borough, 1980: 14). The North Slope Borough programme covered only the mid-Beaufort coastal zone extending to the three-mile limit of the state's jurisdiction and inland to the 200-foot contour. The proposed programme, as prepared by the local government, emphasized "protection of subsistence land and water uses over all other uses ... including oil and gas developments" (U.S., Department of the Interior, USGS, Conservation Division, 1980: 72).

Senior government

Both the federal and state governments have significant roles in offshore oil and gas development — for example, through the Coastal Zone Management Act. In addition to the formal statutes and regulations noted below, other arrangements among federal agencies and between federal and state agencies have also been made. An example is the Joint Federal/State Beaufort Sea Oil and Gas Lease Sale (U.S., Department of the Interior, USGS, Conservation Division, 1980: Appendix 1) and the agreement between the U.S. and the State of Alaska pursuant to the Outer Continental Shelf Lands Act of 1953 and Alaska Statute 38.05.137 (U.S., Department of the Interior, USGS, Conservation Division, 1980: Appendix 2).

Federal

In the United States, the outer continental shelf (OCS) is under federal ownership and jurisdiction. The Outer Continental Shelf Lands Act, which was amended in 1978, and for which there is no exact counterpart in Scotland or Canada, established the basic framework for OCS oil and gas development. It empowers the Secretary of the Interior to administer the leasing of lands to private industry, and to establish rules for proper regulation of mineral development. Within the Department of the Interior (DOI), the Bureau of Land Management (BLM) is responsible for the overall co-ordination of the leasing programme, while the U.S. Geological Survey (USGS) regulates activity on the leased tracts and publishes orders for exploration and development in the OCS (State of Alaska, DCRA, Division of Community Planning, 1978).

Prior to lease sales, BLM divides an area of interest into a grid of numbered tracts, each approximately nine square miles (5760 acres), and then schedules the area for possible lease sale. A preliminary assessment of the area's petroleum potential is prepared jointly by BLM and USGS, while BLM, with the assistance of other federal and state agencies, prepares a draft environmental impact statement (EIS) once specific tracts proposed for leasing have been selected. The public is invited to comment at hearings on the draft EIS. The final EIS is submitted to the Council on Environmental Quality for comment. The Secretary of the Interior must then decide whether and on what terms to hold the lease sale.

Many other key federal agencies have complex, interacting, and somewhat overlapping responsibilities in the OCS development process. For example, the Army Corps of Engineers (Department of the Army) issues permits for all permanent and temporary structures and discharges in navigable waters pursuant to section 404 of the Federal Water Pollution Control Act of 1972 and the River and Harbor Act of 1930. The Corps of Engineers is also responsible for authorizing artificial islands or fixed structures on the outer continental shelf under the Outer Continental Shelf Lands Act and its amendments. Under the Marine Protection, Research, and Sanctuaries Act of 1972, the Corps of Engineers issues permits for ocean dumping outside the limits of the territorial sea. The Environmental Protection Agency (EPA) is responsible for authorizing the discharge of pollutants into navigable waters under section 402 of the Federal Water Pollution Control Act. The Coast Guard (Department of Transportation) has the power to authorize the construction or modification of bridge structures and causeways across navigable waters and to authorize the siting, construction, and operation of deep-water ports. The Outer Continental Shelf Lands Act also grants authority to the Coast Guard to promulgate and enforce regulations for safety-related matters on fixed OCS platforms and drilling vessels.

The United States Fish and Wildlife Service of DOI and the Department of Commerce's National Oceanic and Atmospheric Administration (NOAA) provide recommendations for the protection of fish, wildlife, and natural habitat to permitting agencies such as the Corps of Engineers. NOAA's Office of Coastal Zone Management (OCZM) administers the Coastal Zone Management Act, which provides grants to states for the development and implementation of management programmes to control land and water uses within the coastal zone. Under title 11 of the Marine Protection, Research, and Sanctuaries Act, the Secretary of Commerce may recommend areas as marine sanctuaries to the president, following consultation with other agencies such as DOI and EPA, and may issue regulations applicable to such sanctuaries. Other authority has been granted to the Department of Commerce under the Fishery Conservation and Management Act of 1976, the Marine Mammal Protection Act of 1972, the Endangered Species Act of 1973, the Fur Seal Act of 1966, and the National Ocean Pollution Planning Act of 1978.

State

Six Alaskan state agencies have offshore oil and gas development responsibilities which interact and overlap to varying degrees with federal and local government responsibilities. These state agencies are: Department of Natural Resources (DNR), Department of Environmental Conservation (DEC), Department of Fish and Game (DFG), Department of Transportation and Public Facilities, Department of Community and Regional Affairs (DCRA), and Division of Policy Development and Planning (DPDP) in the Office of the Governor.

The State of Alaska has jurisdiction in the offshore area within the first three miles (five km.), beyond which the federal government has jurisdiction, as established by the Outer Continental Shelf Lands Act. Alaska's Department of Natural Resources has primary responsibility for the management of the state's subsurface energy and mineral resources. It leases offshore lands for extraction of minerals, oil, and gas. It also issues leases and permits for the use of tidelands and submerged lands for purposes other than the extraction of natural resources, as well as permits to modify any body of non-marine surface water (State of Alaska, DCRA, Division of Community Planning, 1979).

The Department of Environmental Conservation issues waste water and solid waste permits, enforces air quality regulations, and regulates tankers in state offshore waters. The Department of Fish and Game issues a variety of permits for activities within designated state game sanctuaries or refuges, and fish and game critical habitat areas.

The Division of Policy Development and Planning within the Office of the Governor includes the state Office of Coastal Management and the State-Federal Clearinghouse, which will be dealt with in greater detail later under the section on co-ordination in chapter five. In addition to being responsible for the Alaska Coastal Management Program (ACMP), the Office of Coastal Management serves as staff for the Alaska Coastal Policy Council (ACPC). The council consists of sixteen members: seven are heads of state agencies and nine are locally elected officials appointed by the governor. The council adopted standards and guidelines which constitute the minimum requirements for district programmes; the council, together with the legislature, must approve all district or local programmes. These local coastal management programmes and the ACMP regulations can influence the entire range of onshore activities, from water requirements to waste disposal and facility siting.

Local or district governments receive funding and technical assistance from the Department of Community and Regional Affairs, both in preparing coastal management programmes and in planning for and managing interrelated onshore and offshore oil and gas developments. This department provides local communities with information on: (1) types of onshore facilities that may be required as a result of offshore developments; (2) the implications of onshore developments for communities; and (3) the role of the state government in setting policy and managing development (David M. Dornbush, 1976). Numerous reports dealing with these and other related matters have been written or commissioned by the department as an aid to local government. A series of five guidebooks aimed at the general public has also been prepared by DCRA on coastal management and how to develop a district programme (State of Alaska, DCRA, Division of Community Planning, 1979).

Industry

In Alaska, and in the United States generally, public oil and gas development corporations do not exist. Active private corporations generally belong to a variety of associations which perform lobbying and research functions. Two national organizations are the American Petroleum Institute (API) and the National Ocean Industries Association (NOIA). The oil and gas companies in Alaska have also formed the Alaska Oil and Gas Association (AOGA). In addition to lobbying in national and state capitals, these groups do research and publish brochures for the general public.

Another industry association has been established in Alaska and is known as ABSORB — Alaskan Beaufort Sea Oil Response Body. This organization consists of companies involved in the Beaufort Sea, and is responsible for producing an oil spill contingency plan which will assist each company in preparing for offshore exploratory drilling activities in the Beaufort Sea Lease Sale Area.

Voluntary organizations

Numerous environmental interest groups participate in public hearings, present briefs, comment on environmental impact statements, and evaluate government programmes. Some of the more active groups in Alaska include: Friends of the Earth, Trustees for Alaska, Greenpeace, the Conservation Foundation, the Alaska Center for the Environment, and the Fairbanks Environmental Center. Some, such as Friends of the Earth, publish regular newsletters and co-ordinate the responses of a number of organizations on various issues and on environmental impact statements. In many cases these groups have in-house legal expertise which allows them to undertake litigation, a practice more common in the U.S. than elsewhere.

Onshore oil development, Alaska, Dan Kowalski-Lightwriting, Inc., Juneau, Alaska

Chapter III: Planning

Introduction

Planning involves an array of activities which permit government and other actors to anticipate and guide development and its effects to varying degrees. Of basic importance is strategic planning, an activity which is generally envisioned as identifying goals and objectives, technical or other means of achieving them, and criteria for judging success. Strategic planning should therefore involve consideration of inventory, land classification, environmentally significant areas, evaluation, and implementation (see Table 1). The planning approaches of various actors in Scotland and Alaska are discussed below, chiefly as a basis for evaluating management effectiveness later in the report.

Strategic planning

Scotland

Local government

Regional government structure plans are a major strategic planning tool in Scotland (G.B., Central Office of Information, 1979; Houston, 1977; Batey and Breheny, 1978). An innovation of the 1960s and 1970s, these structure plans include economic, social, and environmental goals and objectives, and identification of key concerns, as well as the outlines of technical and other programmes to meet these goals and concerns. Theoretically, the plans are reviewed every five years. Every structure plan must be approved by the Secretary of State for Scotland before the relevant local or district plans can be adopted. The secretary can modify structure plans in light of national or regional considerations, conflicts, or controversy. Many different approaches are taken to plan preparation in Scotland and the U.K. as a whole, including some relatively sophisticated system approaches which attempt to integrate various topical or subject plans — such as housing and recreation — with overall regional and district plans (Batey and Breheny, 1978).

Senior government

Major responsibility for strategic planning for North Sea oil and gas developments lies with the Department of Energy and the Cabinet. The Department of Energy can control which North Sea blocks will be opened to exploration, and eventually to production (if sufficient petroleum is found). Through its licensing powers the department can also control which wells will be drilled and under what technical, economic, environmental, and other conditions. The British government and in particular the Department of Energy are, of course, strongly influenced by industry in these matters. Industrial willingness to invest and industrial knowledge and information are very strong strategic planning influences.

Seven rounds of block allocations have been undertaken by the U.K. in the North Sea since 1971. Resource or environmental assessments are not known to have been used in any offshore drilling programme until very recently in the case of the Beatrice field close to shore in Moray Firth in northeast Scotland.

Far more consideration of resource impacts, conflicts, and environmental effects has occurred with regard to land than to sea through local and senior government use of inventories, land classification systems, and other aspects of management. Decisions on the general location and effects of pipelines, oil terminals, and other facilities on land are often influenced by decisions on pipeline locations or other facilities at sea. The extent to which the Scottish Office or other bodies serve as co-ordinating devices among the Department of Energy, other senior departments, industry, and local government is not entirely clear. In a number of cases the statutory planning process has resulted in changes in planned locations of pipelines or other facilities — sometimes for resource and environmental reasons. In a few areas, notably the Shetlands, much anticipatory planning has occurred with considerable impact on the nature, size, location, and other characteristics of coastal petroleum development facilities and their effects.

No comprehensive statement on North Sea petroleum development policy and its economic, social, and environmental implications for Scotland and the U.K. in general has been issued by the Department of Energy or the Cabinet (University of Aberdeen, Department of Political Economy, 1978). The British government has, however, declared that its policy is to develop the North Sea deposits as quickly as possible in order to avoid dependence upon costly foreign petroleum. Balance of payment, investment, entrepreneurial, employment, and other economic and social objectives have been prime objects of British policy. The decision for rapid development was made without much public discussion. It does, however, have considerable support in Scotland. The economic planning and development departments in the Scottish Office have, as a general rule, favoured development and attempted to expedite planning as rapidly as possible.

Early in the 1970s, individuals in industry, local and senior government, and the universities recognized the possibility of substantial undesirable impacts on the Shetlands environment and way of life. It was therefore proposed that a group be created to include representatives of the oil industry and what was then the Zetland Islands Council (now the Shetland Islands Council), along with selected experts from universities and other organizations. The purpose was to discuss the developments and any environmental implications freely and fully. Through this approach, it was hoped that many of the problems associated with harmonizing developments with the Shetlands situation could be identified and resolved before planning applications were submitted (Dunnet, 1980).

The first organization set up to address these goals — the Sullom Voe Environmental Advisory Group (SVEAG) — met regularly over a two-year period (1974-76), and produced a preliminary environmental impact assessment, published in the Shetlands in June 1976 by SVEAG. The EIA essentially examined environmental aspects of a single oil terminal site identified in previous Shetlands-sponsored research and made a series of recommendations on construction issues. Difficulties arose over the composition and operation of SVEAG, mainly because of disagreements over the desirability of underground or above-ground oil storage tanks. For this and other reasons, including the fact that the development process was at the end of the initial planning stage, SVEAG was terminated in the summer of 1976.

Subsequently, SOTEAG was set up with the terms of reference noted in chapter two. It meets approximately six times a year, usually in either the Shetlands or in Aberdeen, to discuss problems which are brought to it by the Shetland Islands Council, the oil industry, or the Sullom Voe Association, and to deal with matters generated by the group itself. SOTEAG has a wide interest in chronic pollution and other biophysical effects of development; it is not directly concerned with resource conflicts or socio-economic matters.

Industry

The basic planning and management stance of developers is opportunistic. Many large corporations have strategic planning units which prepare economic forecasts of opportunities as well as engineering or other technical means for meeting them. Government departments with strong interests in development and close ties to developers often prepare similar studies, frequently in close association with industry. A recent example is a Department of Energy study of a North Sea gas gathering system (G.B., Department of Energy, 1978). This and other reports are frequently narrow in orientation, and focus primarily on economic or technical issues.

Environmental management appears to be given little consideration in the early strategic planning undertaken by the more than thirty large corporations involved in North Sea petroleum development. However, once basic decisions on policy, programmes, and projects have been made — primarily on the basis of their economic implications for the developer — then more attention may be devoted to resource impacts, to conflicts with fishermen and other users, and to route selection and other adjustments required to secure planning permission from government.

Insufficient information is available to give a comprehensive view of approval and implementation planning for the industry as a whole. However, the approach is quite varied. As in other economic growth situations, certain companies engage in speculation; they attempt to gain opportunities by securing control of key sites or other factors. Considerable speculation occurred in the mid-1970s over opportunities to build large platforms, particularly the concrete structures sometimes used for production in deep water. Old and newly created companies were involved, many with little interest in the social, resource, or environmental implications of their enterprises. Senior government perceived the companies' interest as indicative of a high demand and acted through the Offshore Petroleum Development (Scotland) Act 1975, and in other ways, to expedite acquisition and planning procedures in the national interest. The expected demand never materialized.

In one very controversial case, an application for a concrete platform site at Drumbuie in the isolated west highlands was ultimately rejected by the Secretary of State for Scotland after a public inquiry, largely because the proposal involved a scenic National Trust for Scotland property. However, quick permission was given subsequently for a site at nearby Kishorn without recourse to normal planning procedures. Certain social and environmental safeguards were required in this project, but a number of them are not now being met and are unlikely to be enforced by either local or senior government agencies (Baldwin and Baldwin, 1975: 89-90).

Some companies, especially those with a longer-term commitment to Scotland and the U.K. as a whole, have developed relatively comprehensive approval and implementation procedures. An example is British Gas Corporation which has a thirty-person staff in its environmental planning office. The staff consists of biologists, geographers, engineers, planners, and other professionals capable of carefully analyzing the resource, environmental, and other implications of projects. The company uses a system of field liaison officers for public relations and inspection purposes. Through its planning unit, the corporation has developed close contact with the Scottish Development Department and other agencies. Approval and implementation work is conducted in three stages: a provisional stage; a detailed planning stage; and a final planning permission/public consultation stage. Such planning is strongly guided by project-cost constraints.

Voluntary organizations

A principal goal of many voluntary organizations is to influence decision-making on matters relating to birds, soil protection, scenery, or other elements of the landscape and the environment. However, these organizations often have other roles which impinge upon the policy role to varying degrees. An example is the Scottish Wildlife Trust, which manages more than forty wildlife reserves. Its members also undertake habitat surveys, design nature trails, and produce trail booklets.

One of the more active organizations is the National Farmer's Union of Scotland. Among its concerns is the general protection of agricultural resources. The union and other interested groups have therefore pressed for careful construction and inspection procedures, notably for pipelines. The extent to which management and rehabilitation procedures are successful is debatable. Few systematic evaluations appear to have been completed after construction. Looping and other pipeline activities tend to occur in the same general areas. The effect of such modifications is not fully understood, nor is it fully taken into account in initial planning.

The National Trust for Scotland has the important role of preserving places of historic or architectural interest, or natural beauty. It does not perform a major policy evaluation role, although it has been involved in some oil project conflicts. The trust is financially dependent on legacies, donations, and members' subscriptions. It co-operates with government bodies in planning and managing its properties, a number of which have been set up as county parks in co-operation with the Countryside Commission for Scotland. The trust owns about seventy properties covering 80,000 acres. A number of these are located in scenic coastal locations and tend to be treated as environmentally significant areas.

The Royal Society for the Protection of Birds aims to encourage better conservation and protection of wild birds and their habitats. Like the National Trust for Scotland and the Scottish Wildlife Trust, the society is in the land management business, being responsible for seventy nature reserves, of which twenty-three are in Scotland. Unlike these other bodies, however, the society has a major policy, practice, research, and educational thrust. Its objectives include: (1) the detection and reduction of threats to birds and habitats from changing land use or pollution; (2) scientific research; (3) enforcement of bird laws; (4) education work, including film production; and (5) publishing and the provision of advice to schools. The society has a staff of over eighty, of which about two-thirds are professionals. The society has been quite active in the Shetlands, where it has a full-time field officer.

The British Field Studies Council is an independent educational organization that has conducted studies on oil development impacts on the biophysical environment. The council is a registered charity that depends for its income on fees charged for membership, instruction, use of its ten U.K. field centres, and other services. The council employs qualified staff in a wide range of field subjects, and offers courses in plant and animal ecology, geography, geology, land use and planning, archaeology, and art.

The universities of Scotland and other parts of the U.K. represent another type of major voluntary or third-party institution involved in North Sea oil developments. Publication lists for the Scottish coast, the Shetlands, and other affected areas show numerous individual and team research projects by faculty and students. University members are also involved in key organizations such as SOTEAG. Indeed, faculty members have played a leading role in establishing such liaison and co-ordinating groups. University staff have also been employed frequently as consultants. Scottish offshore oil development also has prompted the formation or further growth of study groups or institutes to promote university research on concerns highlighted by oil-induced changes. An example is the Institute for Study of Sparsely Populated Areas at the University of Aberdeen. This institute focuses on interdisciplinary social scientific research on peripheral regions and areas of low population density, many of which are now being heavily impacted by oil development.

Alaska

Local government

In Alaska, local or district coastal management plans are a major strategic planning device. These plans allow the local government to designate or zone the industrial, conservation, or other uses to which various sections of the coast may be put. Once these plans have been approved by the state legislature they have the force of law and all federal, state, and local government agencies, as well as the public, must conform to them. In preparing these plans, the districts must conduct inventories either directly or by using existing information on the natural and cultural resources of the coast.

Natural resource inventories are intended to include consideration of topography, geology, climate, hydrology, soils, vegetation, and habitats. The cultural resource inventory should include consideration of land and water uses; major land and resource ownership and management responsibility; major historic, prehistoric, and archaeological resources; population; and public access (State of Alaska, DCRA, Division of Community Planning, 1979: Vol. 2). The yet-to-be-accepted North Slope Borough coastal management plan includes such information as well as a zoning ordinance developed to regulate onshore oil exploration and development activities.

Senior government

The federal government and the Alaskan state government are both required by law to develop five-year oil and gas leasing schedules, which are important strategic planning devices. These schedules are viewed as providing a measure of certainty to the leasing process, giving all interested groups an opportunity to comment, and permitting local communities to plan and prepare for developments and their effects.

Federal

Under section 18 of the Outer Continental Shelf Lands Act, the Secretary of the Interior is required to prepare a leasing programme consisting of a schedule of proposed offshore lease sales, and specification of the size, timing, and location of leasing activity that will best meet national energy needs for the five-year period following its approval. In addition, the programme must include "estimates of appropriations and staff required to obtain resource information, analyze and interpret exploratory data, conduct environmental studies and supervise operations conducted pursuant to each lease" (U.S., Department of the Interior, BLM, 1980b: 1).

Development of the five-year programme for the leasing period June 1980 to May 1985 began in October 1978, and the programme was finally approved in June 1980. Extensive involvement by state and local governments and environmental and industry groups took place during the almost two-year development period. Information was requested from the governors of affected coastal states and from federal departments such as Justice, Energy, and Commerce. Public meetings were held throughout the country, including meetings with industry and environmental groups. Comments on the draft programme released in March 1979 were taken into consideration in the final decisions on the timing and location of lease sales for the U.S. continental shelf up to May 1985.

Another important U.S. strategic planning initiative is the National Oceanic and Atmospheric Administration's "strategic regional assessment studies of U.S. coastal areas." These assessments are being conducted by the agency's Office of Ocean Resources Coordination and Assessment (ORCA) for the Gulf of Mexico, the east coast, the arctic Alaska region, the Gulf of Alaska to the Oregon coast, and California. In essence, the aim is to develop an ocean and coastal resources management framework for analyzing and evaluating strategies for reducing resource-use conflicts and for protecting the environment (Duke et al., 1980: 3).

The east coast project involves mapping five categories of data for the coastal and ocean zones: (1) physical environment; (2) living environment; (3) species; (4) economic activities; and (5) jurisdictions. The study areas include all coastal counties, and extend seaward to the 200-mile limit of the U.S. fisheries zone. Over 100 maps at a scale of 1:2,000,000 will comprise the data base. In addition, composite maps will be developed. The final project will consist of a data atlas, a data bank, and a final report. The report is to be organized into four main sections:

1. a regional characterization of the biophysical, economic, and jurisdictional regimes of the east coast;
2. a description of the data base and how to use it;
3. application of the data base to at least two problems—the evaluation of alternative locations for major energy facilities ... and the identification of areas that have special protection status; and
4. conclusions and recommendations (U.S., Council on Environmental Quality, and U.S., DOC, NOAA, Office of Coastal Zone Management, Office of Ocean Resources Coordination and Assessment, n.d.: 1).

No comparable senior government strategies planning system has been developed in either Scotland or the Canadian Beaufort area, although in the eastern Canadian Arctic the recent Lancaster Sound inventory, classification, and planning procedures are steps in this direction.

State

In 1978, Alaska's statute on subsurface energy and mineral resources management was amended. Among the changes was the requirement that the Commissioner of Natural Resources submit to the legislature a five-year schedule outlining state land leasing for oil and gas exploration. The first of these five-year lease plans was submitted in January 1979, and two sales on that schedule have been held. One of these was the Joint Federal/State Beaufort Sea Oil and Gas Lease Sale.

A major purpose of the state five-year schedules is "to provide a plan to facilitate the orderly assessment and development of Alaska petroleum resources." A fixed and predictable leasing programme is viewed as being advantageous for local citizens and government, environmental groups, and industry:

> An established leasing schedule permits the public to comment on the planned areas of leasing, allows business and labour to plan employment or business opportunities, permits the petroleum industry to allocate a portion of its resources to Alaska petroleum exploration and development, and provides local, state and federal governments with adequate time for presale evaluation, assessment and mitigation of impacts (State of Alaska, DNR, Office of the Commissioner, 1980: 2).

In conjunction with the January 1979 five-year schedule, Alaska's governor issued Administrative Order 52 which established the Agency Advisory Committee on Oil and Gas Leasing to ensure that before each proposed lease sale, its social, environmental, and economic impacts are analyzed. This committee consists of the commissioners of eight state departments, a member of the governor's office, and a member of the local community affected. Both formal hearings and informal consultations are provided by the committee to allow public, local government, industry, and other groups to express their views.

In preparing its five-year schedule, the state has aimed at co-ordination with the federal OCS schedule. This approach has a number of advantages, including: "the generation of industry interest in an area, exchange of geological data, and the ability to use the environmental information and studies generated by the Federal Environmental Impact Statement process" (State of Alaska, DNR, Office of the Commissioner, 1980: 9). In concluding this section, it should also be noted that the Alaska Coastal Management Program is an important overall strategic planning tool.

Inventories

Scotland

Municipal and senior governments collect and use much demographic, economic, social, and environmental information. Some agencies have, or are developing, large multi-factor and computer-based data banks (Batey and Breheny, 1978). A variety of land use and other inventories have been conducted in Scotland and the United Kingdom. Many of these have been directly or indirectly supported by senior government departments — for example, the nation-wide land use surveys.

The Nature Conservancy Council and the Countryside Commission for Scotland have mapped many areas of special scenic or environmental significance and about twenty-five per cent of the United Kingdom is now allocated to national parks, areas of outstanding national beauty, sites of special scientific interest (SSSI), and other appropriate designations. However, no overview studies seem to be available on the influence that these designations have had on the location of pipelines or other facilities, or on amelioration of construction or operational effects through special management methods.

In contrast to the situation on land, a relatively small amount of information is available on marine resources and environment. No inventories comparable to the comprehensive land use surveys have been completed. Hydrographic, fish, bird, and other surveys are incomplete for much of the U.K. coastal waters. The lack of such data limits knowledge of short- or long-term damage to fishing grounds or other key resource sites and inhibits effective planning on a multiple-use basis.

Industry relies on existing government, university, and other inventories for wildlife, soils, and other information in project planning. Additional research is usually conducted in-house or by consultants, especially as planning becomes more detailed and more large-scale information of the type not likely to have been of interest to other parties is required.

Many of the voluntary organizations have been and are involved in inventory, land classification, and related activities. The universities are the source of many of the concepts and methods used in such work. The British Field Studies Council has conducted inventory work on oil pollution at Sullom Voe and at other places in Scotland and the rest of the U.K. (Baker, 1977; Baker and Hisock, 1980). This inventory work is part of the monitoring activities that are a focus of the council's research programme. The council is participating in the SOTEAG monitoring programme at Sullom Voe. Its approach is based largely upon extensive experience in taking inventory, monitoring, and analyzing oil spills and their effects at the large oil port and refinery at Milford Haven, Wales.

Alaska

As noted earlier, Alaskan districts are required to collect information on onshore and offshore natural and cultural resources in preparing their local coastal management plans. Much of this information is available through various state departments. For example, the Department of Natural Resources has produced a land use atlas, and an inventory of recreation and heritage resources of the coastline. The Department of Fish and Game has produced maps to assist in delimiting the biophysical boundaries for the coastal zone. However, it is often necessary for the district, on the basis of information obtained from the state, to conduct more-detailed studies.

The federal Bureau of Land Management (BLM) has established a socio-economic studies programme within the Alaska Outer Continental Shelf Office. This programme "is a multi-year research effort which attempts to predict and evaluate the effects of Alaska OCS petroleum development upon the physical, social, and economic environments within the state." There are three main research components:

1. documentation of the nature, structure, and processes of the various geographic units and interactions among them;
2. preparation of development scenarios on the location, nature, and timing of future OCS petroleum development events and related activities; and
3. translation of future oil and gas development events into quantities and forces acting on the geographic units (Dames and Moore, 1978).

As part of the Socio-economic Studies Program (SESP), a series of comprehensive baseline studies of the Beaufort Sea region and communities was undertaken in 1977. Inventory information was compiled on land use, land status, transportation, and utilities (U.S., Department of the Interior, BLM, Alaska OCS Office, 1977), as well as current social, economic, physical, and demographic conditions.

In addition, the Outer Continental Shelf Environmental Studies Program has been established by the federal Bureau of Land Management, with a separate Alaskan programme which is managed by NOAA. Under this programme, research on arctic physical and biological resources is undertaken; this includes basic survey and inventory work. One of the major purposes of this programme is to provide environmental information to assist BLM in making lease sale decisions.

Land classification

Scotland

A wide variety of land classification systems are available, ranging from single-factor, colour-coded maps, to use of a system of map overlays. Opinion varies as to the extent to which the various systems are used by local governments. Complex computer-based maps synthesizing much biophysical and land use information have recently been produced for the Shetlands. A range of land and resource classification systems are used on land, but these are not common for marine and coastal waters. Little original work on classification systems is known to have been undertaken by industry, which relies upon the Nature Conservancy Council (NCC) and other government and university groups for methods and data. Sensitive areas identification was included in the preliminary EIA for the Sullom Voe oil terminal. SOTEAG has produced information which has been used in land classification or in sensitive areas identification by NCC and other groups. SOTEAG remains interested in utilizing such work in its monitoring, assessment, and advisory role.

Especially significant areas

Scotland

Many municipal governments receive information on wildlife and other significant areas from senior government bodies such as the Nature Conservancy Council, which also identifies sites of special scientific interest that the municipalities must consider in granting planning permission. Little environmentally significant area mapping seems to have been undertaken by the municipalities themselves, although the Shetland Islands Council used this basic approach in site selection, development, and environmental planning — notably in the coastal zone. This approach is also being used in detailed planning and management of the Sullom Voe-Yell Sound area near the new oil terminal, chiefly in the context of oil spill hazard.

Large-scale identification of especially significant areas is uncommon in offshore waters, although major fish spawning areas are often known to fishermen. The Nature Conservancy Council and other bodies are attempting to remedy this situation — for example, by mapping key winter staging and other critical areas for birds in U.K. coastal waters. Concern for seabirds has risen sharply as a result of large bird kills through oil spills, the discharge of tanker ballast, and other influences.

Alaska

The Office of Coastal Management has prepared a volume on proposed special management areas for use by local, state, and federal government agencies. The proposed areas are considered suitable for approval as state parks, geophysical hazard areas, or other special management types under the Alaska Coastal Management Program (State of Alaska, DNR, 1980: iii). In developing local coastal management plans, coastal resource districts must designate and develop management policies for areas which merit special attention within their district coastal area boundaries (State of Alaska, Office of the Governor, DPDP, Office of Coastal Management, and U.S., DOC, NOAA, Office of Coastal Zone Management, 1979:93).

Evaluation

Scotland

Local government

General methods of evaluation have been part of the Scottish municipal government decision-making system for decades. However, use of specialized benefit/cost (B/C), environmental impact assessment (EIA), social impact assessment (SIA), and other evaluation methodologies are still uncommon in Scotland, although district councils now request them from the proponents of new projects. EIAs in particular are being recognized as valuable, and a procedures handbook has been published by senior government for use by local governments (Clark et al., 1976).

In the Shetlands, a variety of evaluation studies, used as a basis for making decisions on oil terminal and other developments, are generally considered to have been effective in reducing development impacts. In contrast to the North American situation, careful landscape and architectural design analyses have been completed for several petroleum projects in Scotland and other parts of the U.K. in an attempt to fit new structures aesthetically into the environment.

Senior government

Evaluation is a wide-ranging activity in the oil development field, but is uneven in its distribution and character among senior government agencies and activities. Senior government did support comprehensive studies of recreation and tourism patterns prior to the onset of major petroleum projects on the Scottish mainland (University of Edinburgh, Tourism and Recreation Research Unit, 1976). These studies included strategic and detailed planning recommendations and were reportedly quite useful to some regional and district authorities, although no follow-up research has apparently been undertaken on the studies' effects. Some use of systematic evaluation techniques has occurred, especially in recent years. The application of such methods is not, however, a consistent policy tool in Scotland and the U.K. as a whole — as it is in the U.S. with its federal requirement under the National Environmental Policy Act (NEPA) of 1969 for environmental impact assessment of national policies, programmes, and projects likely to have a significant impact on the social and biophysical environment.

In particular, no systematic use of EIA or comparable assessment methods appears to be present in the transport field — for roads, ports, and air-fields. Thus major oil-induced changes in the route, size, and character of the A9 highway between Inverness and the Glasgow-Edinburgh area were apparently made without any evaluation of their economic or other impacts on the land or upon by-passed settlements of central Scotland. A number of these centres had previously received tourist development assistance from the Scottish Highlands and Islands Development Board.

Some economic and social evaluations have been supported by senior government. These studies can provide a better conceptual understanding of the nature of oil development and its effects. An example is the concept of the oil cycle and its phases as presented by a University of Aberdeen study team. This concept and associated findings are intended to provide a framework whereby local authorities and other interested parties might plan for potential social effects of petroleum development. The framework also has been used by the Aberdeen study team as an organizational tool in case studies of impacts in the Shetlands and other parts of Scotland.

In the view of the Aberdeen group, the oil cycle consists of three phases: (1) exploration and appraisal; (2) development; and (3) production. During these three phases, different types and levels of offshore and onshore activity come into play (G.B., Scottish Office, SDD, Central Research Unit, n.d.).

Phase I of the oil cycle — exploration — involves identifying geological structures and assessing economic viability. Offshore activity tends to build rapidly, while onshore activity is generally limited to service facilities. A speculative atmosphere prevails offshore and uncertainty about the speed, scale, and duration of activity prevails onshore. This phase characterized the northern North Sea from about 1967 to 1975.

Phase II — development — occurs when a field is found to be economically viable. This phase marks the peak of onshore activity, with much concern arising about construction, production, transportation, and primary oil and gas processing. Key decisions are taken — often quite rapidly — on the type and location of onshore facilities. These decisions largely set the development and land use patterns for the rest of the cycle. Unknowns of many kinds make the length and character of this phase uncertain. Among the key influences are technical difficulties, market uncertainties, licensing, and other policy procedures. The development phase is the largest in scale and in terms of its employment, economic, and social effects. Employment in petroleum-related projects is in the thousands as opposed to hundreds in the production phase. This is the critical stage that is now being entered in the Canadian Beaufort Sea.

Phase III — production — has been experienced to a limited extent in Scotland. This phase also has many uncertainties with regard to scale, nature, and effects. In later years, as reserves become exhausted, both offshore and onshore facilities decline. There is doubt as to when and how this will occur, with economic conditions, technical advances, and company and government policies all being important influences.

The University of Aberdeen study team has not tested the oil cycle concept against the resource and environmental effects of petroleum development. It may be that the environmental impact pattern will differ somewhat from the socio-economic pattern. For example, certain environmental effects could be relatively large in the production phase if long-term chronic pollution proves to be especially damaging to life.

Industry

Economic and technical evaluations are apparently conducted frequently by industries involved in North Sea petroleum development, although few comprehensive benefit/cost analyses of various planning alternatives have been released by these companies. Social impact assessments appear to be undertaken rarely, at least as separate exercises in the North American sense. Certain social effects — for example, on housing, transport, or other land use aspects of development — are, however, included in many environmental impact assessments. Many developers are not supportive of EIA, although some, such as British Gas Corporation, recognized the value of this type of analysis more than ten years ago (Dean, 1979; Dean, n.d.; Dean and Graham, 1979; Dean, Graham and Ward, 1979).

The objectives of the British Gas Corporation EIA system are: (1) to produce a better-designed development; (2) to improve relations among developers, planning authorities, and local communities; and (3) to obtain planning consent within the time-scale specified in the corporate plan. According to British Gas, a principal value of EIA is in showing the public that all relevant factors have been considered in selecting the location for a facility. Another associated value is that the overall time to bring a project to the implementation stage is reduced. Department of the Environment surveys conducted in the late 1970s indicate that the average time to secure a planning consent (termed outline planning permission) for a major industrial project on a "greenfield site" in an area not zoned or designated for industrial use was not less than two years from date of submission of the application. In contrast, since 1966 British Gas, through environmental impact assessment, has been able to process fifty major projects within an average time period of five to six months.

Not surprisingly, British Gas has also found economic advantages in EIA, if only in terms of time and related cost savings. The total cost of the fifty projects has been £300 million, resulting in a company-estimated inflation saving of some ten per cent, or £30 million over the eighteen-month, shorter-than-average planning period. During this time, British Gas reportedly spent an estimated 2.5 per cent of total project costs (£7 million) on environmental planning (Dean, Graham and Ward, 1979: 37).

Another type of analysis often carried out by British Gas and other corporations is visual impact analysis, which can be part of an EIA. British Gas stresses that architects and other designers should be involved from the earliest stage in facilities planning in order to secure the full benefit of their skills and avoid a cosmetic approach. Various innovative computer-based methods for scaling drawings of proposed facilities into actual photographs of a site have been developed as an aid to visual analysis.

Voluntary organizations

As noted earlier, universities and other non-government organizations have participated in some of the environmental impact and other evaluations of oil and gas development in Scotland. Relatively few evaluations of the efficiency or effectiveness of policies, management guidelines, or planning conditions have been undertaken, so that it is difficult to determine how well they are working and whether they should be applied elsewhere. A group at the University of Aberdeen known as the Project Appraisal for Development Control (PADC) is investigating the links between project appraisal and plan-making. A central concern is to determine how the problems identified in the approval process were handled in planning and how the processes of evaluation and planning can be improved in the interests of better management.

A University of Aberdeen ecological study of Moray Firth development shows how useful studies of the effects of guidelines and other management tools can be in evaluating policy and practice. The study essentially consists of an examination of changes in agricultural land use and other activities in the Moray Firth area between 1966 and 1980 (University of Aberdeen, Department of Geography, 1980). The study reveals that senior government attempts to protect prime agricultural land against oil-induced impacts have not been very successful. Despite the directive issued by the Secretary of State for Scotland in 1977 (subsequently included in the National Planning Guidelines) that there should be a presumption against taking high quality farmland from agriculture, it is clear that such land has been taken, and in quantity, in the Moray Firth area (see Table 3). A considerable amount of good agricultural land has also been zoned by the local council for further housing, and other oil-induced developments.

The reasons for the failure to implement the agriculture land protection policy are not entirely clear, but seem to consist of an amalgam of the following: (1) insufficiently strong commitment by senior government; (2) a disposition on the part of some local councils and other agencies to favour industrial and other development; (3) the location of good farmland near villages and other centres selected for housing expansion; and (4) the lack of an overall knowledge of the extent of the loss of good agricultural land in the Moray Firth area.

Alaska

The federal National Environmental Policy Act (NEPA) requires that an environmental impact statement (EIS) be prepared as part of the review and approval process of major federal activities. An EIS was prepared for the proposed federal Five-year OCS Leasing Schedule and for the Alaska Coastal Management Program. There are a number of stages in the development of an EIS, each of which involves some form of public participation, including suggestions for issues to be addressed in the EIS, written comments on the

Table 3

Acreages of agricultural land passing to non-agricultural uses: 1969-1978, Moray Firth sub-region

Settlement	1969 - 1978 (Oct.)				
	Class 1	Class 2	Class 3	Other	Total
Alness	–	59.04	296.17	42.84	398.05
Ardersier	–	–	–	165.25	165.25
Balintore	–	–	–	19.70	19.70
Beauly	–	2.70	–	–	2.70
Charleston	–	–	2.88	–	2.88
Cononbridge	–	–	59.51	–	59.51
Cromarty	–	–	–	–	–
Culbokie	–	–	10.08	5.04	15.12
Dingwall	–	15.48	10.62	–	26.10
Evanton	–	9.90	12.63	2.67	25.20
Hill of Fearn	–	6.84	–	–	6.84
Invergordon	62.20	224.34	105.84	24.30	416.68
Inverness	149.84	70.00	203.76	32.00	455.60
Kirkhill	7.82	–	–	2.34	10.16
Maryburgh	–	6.48	6.48	21.78	34.74
Milton	–	29.88	–	–	29.88
Muir or Ord	–	–	24.65	35.65	60.30
Munlochy	–	–	–	–	–
Nairn	–	21.60	140.00	–	161.60
Nigg	–	–	–	44.46	44.46
Rosemarkie-Fortrose	3.24	2.88	–	–	6.12
Smithton-Balloch-Culloden	3.60	5.04	121.14	–	129.78
Taln	–	13.76	206.15	–	219.91
Totals	226.70	446.34	1059.91	397.53	2290.50

Source: University of Aberdeen, Department of Geography, 1980

draft EIS, representations at hearings on the EIS, and comments on the final EIS.

The U.S. Geological Survey conducts a technology assessment programme as a result of the 1978 amendments to the Outer Continental Shelf Lands Act, which requires that USGS ensure the use of best available and safest technology (BAST) on all new drilling and production operations (U.S., DOI, USGS, Office of Deputy Division Chief for Offshore Minerals Regulation, 1980: 2). Through the BAST program, USGS will review its OCS standards and orders, and where necessary, make changes to them.

The Socio-economic Studies Program and the Outer Continental Shelf Environmental Assessment Program (OCSEAP) of the federal Bureau of Land Management in Alaska involve impact assessment research. The SESP for the Beaufort Sea area includes potential offshore petroleum development impacts on socio-economic and physical systems, transportation systems, the natural physical environment, and the socio-cultural system (U.S., Department of the Interior, Bureau of Land Management, 1979a: Vol. 1, 46).

At the state level, under Administrative Order 52 the Agency Advisory Committee on Oil and Gas Leasing is responsible for preparing a social, economic, and environmental analysis (SEEA) prior to each lease sale. The first such SEEA was prepared in June 1980 for a sale in the Upper Cook Inlet. Major projects requiring such a review must be designated by the governor. The following factors are to be considered in the review: economic; community well-being; fiscal; resource utilization; environmental quality; alternatives; and mitigation measures. In addition, the Department of Community and Regional Affairs has purchased a copy of a relevant analytical model (SCOOP) from the Scottish Office. This programme was developed by the North Sea oil research team at the University of Aberdeen to calculate the economic impact of North Sea oil on the Scottish economy. This model provides forecasts on employment, oil and gas production, equipment needed, and land requirements on both a state and regional basis (Trimble, 1977).

Voluntary organizations such as Friends of the Earth, Greenpeace, Natural Resources Defense Council, Trustees for Alaska, and others provide an important evaluating function by commenting on environmental impact statements. In instances where such groups have reason to believe that their concerns have not been adequately addressed by the respective government agencies, litigation is often pursued. In the case of the EIS for the Joint Federal/State Beaufort Sea Oil and Gas Lease Sale, the North Slope Borough and the National Wildlife Federation filed law suits on the basis that the National Environmental Policy Act had been violated because cumulative impacts, alternative lease stipulations, and other possible management schemes had not been addressed adequately. The court agreed and a draft supplement to the EIS was prepared (U.S., Department of the Interior, BLM, 1980a: 2).

Residential development on the coast near the Sullom Voe oil terminal, Shetland Islands, September 1980, J.G. Nelson

Chapter IV: Implementation

Introduction

Implementation involves arrangements for approval, construction, and operation (see Table 1). Many resource and environmental management processes are involved in each of these basic activities. In the following discussion, attempts have been made to focus on what seem to be the major ones: inspection, surveillance, monitoring, enforcement, and management of modifications during construction and operations. Distinctions between these processes can be difficult to make and are discussed to some degree in the questionnaire in Appendix 1. It is important to emphasize that these management processes interact with one another as well as with other processes considered previously in the planning section of this report. Such is the nature of an effective comprehensive management system.

Approvals

Although no formal classification of the basic ways of securing approval for exploration or other land use activities has been developed, two broad types seem to be recognized. The first is adversarial in nature and involves a relatively comprehensive examination of activities, technology, and environmental and social impacts *prior to* giving permission to proceed. It is probably best exemplified in the U.S. — for example, in the securing of permission to drill on the Alaskan North Slope. The second is a negotiation process and generally involves close consultation between industry and government approval bodies throughout planning. This latter approach appears to have been more commonly used in Canada and the U.K., although it has been attracting more attention lately in the U.S. where it is becoming known as "negotiated development." In discussing approval processes in the context of these two broad types, mention will be made of pre-hearing, hearing, formal case record, conditions or stipulations, appeals, and other procedures considered desirable by the Conservation Foundation as a result of its world-wide land use regulation and planning studies (Noble, Banta, and Rosenberg, 1977; Rivkin, 1977).

Scotland

Negotiated development has long been the rule in granting planning permission in Scotland and the U.K. in general. Large-scale developers generally conduct informal discussions with senior and local government planning officers prior to submitting a proposal. Application is often made for "outline planning permission," which addresses significant elements, principles, and responsibilities, leaving many details to be worked out by governments and industry as the project develops.

Pre-hearings in which the government or other interested parties are given an opportunity to comment are uncommon, as is the keeping of any formal case record on negotiations. A form of hearing occurs when municipal councils debate an application for approval. Direct access of the developer or the public to such council meetings is inconsistent; some councils allow proposals to be commented upon in open meetings by interested parties, while others do not.

Permission can be granted with conditions, but there have been difficulties in this regard, with many water quality or other environmental concerns and stipulations being viewed as outside the jurisdiction of local government and in the domain of water boards or senior government agencies. Councils have increasingly used a special clause (section 50) in the Town and Country Planning (Scotland) Act 1972 to negotiate and impose conditions. However, agreements can be private ones between the council and the developer, as can data on water quality levels or other environmental measures of concern to interested parties, including other resource users.

Project decisions by district councils can be appealed by developers to the Secretary of State for Scotland. In this and other cases — for example, when controversy is high — the secretary can provide for a public inquiry in which an independent reporter hears evidence and makes a recommendation to the secretary who is, however, not bound by the report. The secretary's decision can, in certain limited circumstances, be appealed to the Court of Session, which can direct reconsideration on procedural or other grounds.

More research is necessary on the approval process for publicly owned or controlled agencies such as the National Coal Board or the British National Oil Corporation. Many of these agencies are statutory undertakers in the sense that their role and the management of their effects is provided for in separate legislation. These agencies are often said to be held to the same general management procedures and principles as private undertakers or developers, including the public inquiry process. On the other hand, it is said that such agencies are sometimes not subjected to the same scrutiny as major private companies. An example is the current and continuing controversy in Scotland and the U.K. as a whole over nuclear energy policy, programmes, and projects. A number of large-scale public inquiries have also been held about airport extensions and other major public enterprises. One result has been the contention that under such circumstances public funds should be provided to certain third-party intervenors on the grounds that senior government is compromised when it is both the agent and the judge of change.

British companies such as British Gas Corporation reportedly try to identify "no go" areas prior to project proposals through inquiries and general surveys intended to locate environmentally significant areas as well as any other major constraints. Once a preferred location has been identified, work enters the detailed planning stage, where, after consultation with statutory and non-statutory groups, as well as further research and refinement, a specific proposal is presented to the public for discussion.

Industry tends to plan for likely modifications in projects — for example, through looping or other means of increasing pipeline capacity, or by acquisition of large sites to permit future growth in plant size as the opportunity arises. These future possibilities are not, however, generally analyzed carefully in EIAs or in the approval process. Attempts are being made to provide a more effective planning context for such modifications — for example, by requiring new planning permission for likely major changes in an approved enterprise.

No detailed information is available on the involvement of voluntary organizations in the approval process, although much of the participation appears to occur at the public inquiry or later stages in planning. Some of the funding and other barriers to further participation have been noted previously.

Alaska

The approval process in Alaska is a more open one than in Scotland. Although most of the offshore permits and approvals are issued by the U.S. federal government, state and local government agencies have numerous opportunities to participate in such decisions.

Of major significance is the consistency provision of the Coastal Zone Management Act. This provision requires that once a state has an approved coastal management programme, all federal actions must be consistent with that programme. The consistency decision is made by a designated agency of the state. In addition to being an approval mechanism, it is also a co-ordination device.

It has been contended that the consistency certificate determination provides the best opportunity for state participation in the federal OCS decision-making process. For states with approved coastal zone management programmes, it allows a reasonable amount of time for review of exploration plan activities (three months, with a possible extension to six months if a reason for delay can be given). It also gives states a chance to develop operating conditions that will bring the activities into compliance with the objectives of the Coastal Zone Management Program (New England River Basins Commission, 1980b: 3).

On the other hand, a major project or significant action in the coastal zone requires multiple government authorizations. Each of these authorizations can require a consistency determination. The State of Alaska and the Alaska Coastal Policy Council are currently considering the best means for co-ordinating these multiple consistency determinations so that only one will be required for a particular project. The extent to which efficiency measures will be implemented remains to be seen. Certainly, however, there is a strong commitment on the part of the State of Alaska to local government and public participation in these decisions.

The National Environmental Policy Act process of public review of environmental impact statements is well-established. Extensive reviews and consultations on the EIS for the Beaufort Sea Lease Sale Area resulted in numerous stipulations or conditions being issued on the leases, many relating specifically to environmental concerns. There were twelve federal stipulations, the first nine of which included provisions for archaeological, biological, and environmental protection and mitigation measures (U.S., Department of the Interior, Bureau of Land Management, 1980a).

Construction and operation

Once approval has been granted and construction begins, surveillance and inspection mechanisms are necessary to ensure compliance with any approval conditions for the management of social and environmental impacts. Monitoring and enforcement procedures are also required. Technical, economic, and other changes may prompt future modifications to conditions, stipulations, or understandings. A certain amount of flexibility may be necessary in the management system for desirable changes to be made.

Scotland

It has not been possible in this report to identify clearly which surveillance and inspection tasks in land use, resource, and environmental management are the responsibility of developers, or of local or senior governments. Municipal health and environmental control officers have certain tasks which seem to overlap with those of the Alkali Inspectorate or other senior government units. Arrangements seem to differ from area to area. More research on these arrangements and their efficiency and effectiveness would be helpful. Monitoring of environmental factors, such as water quality and noise levels, is now widely recognized as inadequate in Scotland, whether conducted by government or industry (O'Riordan, 1979). Few people seem to be assigned such tasks, and often, insufficient data is available to make judgements on the effectiveness of systems.

Strict enforcement of environmental conditions and controls is not characteristic of Scotland for several reasons. A fundamental one is that the negotiated development approach tends to work against a decision to confront and enforce. In other cases, data needed for enforcement are not available. Also, enforcement is discretionary on the part of local authorities, who tend to be aware of political difficulties arising from widespread positive views towards rapid oil development.

Modifications, including provisions for rehabilitation following completion of operations, present a major problem for local authorities. Furthermore, changes are made frequently in the original proposal submitted to planning authorities. Obstacles to preventing unwanted modifications include the enforcement constraints noted above, as well as actual or contended changes in the economic environment in which a developer operates. Another problem has been the absence in original planning permission requests of statements on how any modifications are to be decided upon. Local authorities also have had difficulty in securing bonds or other financial guarantees, largely because of the nature of the Town and Country Planning (Scotland) Act 1972.

Senior government

Some senior government agencies do not undertake environmental surveillance and inspection on a regular systematic basis for several reasons. One is an apparent failure to recognize that this may be a very important function in a period of rapid industrial development, even though it has not been perceived as significant in the past, when the pace was slower and there appeared to be fewer uncertainties. Another is insufficiently trained manpower, which is due in part to recent budget reductions by senior government.

A further reason for the lack of senior government environmental surveillance is the negative effect of split jurisdictions and responsibilities among government agencies and levels. An example is the surveillance and inspection of sites of special scientific interest. These areas are identified by the Nature Conservancy Council as having wildlife, plant, or other resources of special value for aesthetic, educational, and other public purposes. Local authorities are notified of their existence and upon designation are required to consult the council on any proposed land use changes. The council can then make recommendations as to the conditions which should apply to any planning permission issued by the local authority. The authority may accept such conditions. But it is not clear what the respective role of the local authority or the council is in ensuring that the conditions are met by developers. Occasional inspections by the council have revealed failures to follow recommended procedures.

Surveillance and inspection activities grade into longer-term monitoring of developments and their resource and environmental effects. As noted earlier in discussion of local government, there is now widespread concern about the need for monitoring in the U.K. Many project approvals are granted in the face of known uncertainties about the effects of certain practices on pre-existing economic and social activities as well as on plants, water quality, and other aspects of the environment. Oil developments often tend to occur in isolated areas where insufficient research has been conducted to provide the understanding necessary to make reasonably accurate long-term predictions about the impacts of changes. Knowledge gaps in more densely settled areas present similar difficulties. Recognition of such uncertainties in many types of development in different parts of the world has led to recommendations for a more adaptive style of management (Holling, 1978).

In oil and other developments, various conditions may be applied in an attempt to control the possibly negative effects of unknowns. Certain water quality or other standards may be set to prevent toxic and other impacts on fish and other organisms. Certain equipment may be required to ensure that standards are met, with provision for changes if the technology does not prove effective. Fines, tax incentives, educational programmes, special zoning, or other procedures may be introduced to control pollution or other effects of development.

Approvals have often been granted on the basis of such conditions, but without adequate thought for their incorporation into efficient and effective management systems. In some cases, little or no provision is made for monitoring the social or biophysical environment to see if the conditions are working and, if not, to recommend what should be done. Is the control standard too high? Is the equipment inadequate?

Monitoring makes possible a continuous process of revision of the initial impact assessment as well as control measures. Some major issues remain with regard to monitoring. What biophysical and social elements and processes are to be monitored, why, and who is to decide? How are the monitoring systems to be managed? Should there be a central information unit and study group? Will this raise major information access issues? Who will pay? In this context, senior government and industry are offering some support to relatively independent monitoring groups such as SOTEAG and the Oil Pollution Research Unit of the British Field Studies Council. Their data generally seems to be available to all parties and are often presented at conferences and published in reports or technical journals.

SOTEAG

SOTEAG appears to have no regular role in Sullom Voe surveillance and inspection activities. These are the responsibility of the companies and government, although SOTEAG could be called upon to advise on such programmes, especially as they grade into monitoring — a principal interest of this group.

The objectives or terms of reference of the SOTEAG monitoring programme have been presented previously. The stage was set for SOTEAG through inventory and baseline research by SVEAG in the mid-1970s. There are, however, some serious difficulties in monitoring at sites such as Sullom Voe because of the varied nature and impact of contaminants and because of inadequate knowledge about the effects of hydrocarbons on the biota (Foxton, 1979). Emission standards are enforced for concentrations of oily water effluents discharged from land and offshore installations, but it is not yet known what constitutes a "safe" environmental concentration; in other words what level of contaminants will have what effect on which biota, and to what degree. Pollution effects due to normal operations of the Sullom Voe terminal have been expected to be chronic and long-term rather than acute (perhaps as the result of a major oil spill). Effects have been envisioned as subtle, cumulative, and not easily detected until they become lethal. In attempts to meet these challenges, SOTEAG has designed a monitoring programme based on the measurement of a limited number of biological variables selected for their relative sensitivity to pollution and sampled over a wide area or gradient having the terminal as its focus.

The monitoring programme is a flexible one, as unforeseen difficulties can easily arise. A major heavy bunker oil spill in Sullom Voe has already caused widespread pollution and complicated the monitoring programme, which was keyed to a measurement system radiating out from the plant. Changes have also been made in ballast water regulations for entering vessels with the result that tankers come into the voe to take on cargo and discharge ballast from other ports and places. This ballast water may contain heavy metals or other pollutants of foreign origin, the effects of which are unknown, and which the monitoring programme was not designed to deal with.

Industry

Many major companies do have environmental planning units, with some personnel usually assigned on a part-time basis to surveillance, inspection, monitoring, and enforcement duties during construction. Issues include inadequate numbers of qualified inspection officers and difficulties in implementing and enforcing control procedures among the many subcontractors and smaller companies involved in a large project.

Problems with offshore exploration, construction, and development in the 1970s led to a major review of governmental and industrial responsibilities for managing accidental oil spills and associated pollution at sea (G.B., Department of the Environment, 1976). The review was conducted by an inter-departmental group of civil servants that made a series of recommendations, including: local and port authorities should be responsible for spills within one mile of shore; and the Department of Trade (and Coastguard Service) should continue to be responsible for dealing with tanker spills at sea. While government or some other group might organize a spill response, the cost of any action should be borne by industry in accordance with the "polluter pays" principle; this should involve not only direct costs, but also overhead of the clean-up organization and depreciation in capital equipment.

The inter-departmental group recommended that industry should be responsible for "cleaning up spillages" during drilling and other offshore oil development operations. The responsibility should shift to the Department of Trade only when there is high risk to the public interest because of inadequate or injudicious response by the operator, or when there is a threat to the coast. Who would decide when and how to undertake such a shift is not apparent from the report of the group.

A number of recommendations were also directed to offshore operators with regard to the implementation and enforcement of oil pollution control programmes. The operators were to develop an action plan for each installation, to ensure adequate vessels were available for control, and to develop control centres and back-up resources at each readily accessible port.

Plans were to be kept up to date. It was also recommended that there be close co-ordination with other companies and with the Principal Marine Officer (Coastguard Service) acting in various offshore regions on behalf of the Department of Trade.

On 28 July 1977, the Minister of State of the Department of the Environment issued a statement in response to the report on accidental oil pollution at sea, after having considered submissions by interested parties. The statement reveals that the government has basically adopted the recommendations in the report, although some difficulties clearly remain — notably the capacity of local government to respond to major oil spills, and potentially inadequate funding for important central government agencies such as the Department of the Environment. This department must, for example, maintain stockpiles of specialized oil spill equipment by off-setting savings within its existing programme (G.B., Department of the Environment, 1977).

Voluntary organizations

While voluntary organizations are not known to be involved significantly in surveillance and inspection activities, the Royal Society for the Protection of Birds and other major organizations are now employing field officers to participate in as many phases of decision-making as possible.

Some of the most interesting monitoring work has been done by the British Field Studies Council and associated organizations such as harbour authorities at Milford Haven, Sullom Voe, and other major oil ports. The oil spill statistics gathered at Milford Haven over the last twenty years are among the most comprehensive available. The Milford Haven harbour authority has played a major role in collecting these statistics and in pollution control activities. This authority is an unusual one. The act of Parliament which created it lays down that "it shall have regard to the desirability of preserving national beauty, of preserving flora and fauna, and ensuring the easy passage of salmon and sea trout." Its main duty is to take such steps as may be necessary "to maintain, improve, protect and regulate navigation and, in particular, the deep water facilities." The authority conducts its development and conservation mandate under the official title of the Milford Haven Conservancy Board.

In summarizing the data available on oil spills in Milford Haven up to 1974, the harbour master noted that:

1. Over 35,000 tankers have been handled in 14 years, involving perhaps nearly 100,000 movements, and 704 pollutions have occurred (1.95 spills per hundred ships);
2. Of these 704 pollutions, 612 have been under 160 gal., and of the remaining 92 not more than 20 have been above 5 tons (1 per 2000 ships handled);

3. The frequency of pollutions is decreasing, both absolutely and, more significantly, relative to the number of ships handled;
4. Over 450 million tons of oil have been handled. Excluding six major spills involving damage or grounding, about 250 tons have been spilt (about half a ton per million tons handled);
5. Of the six major spills which have occurred, two amounted to approximately 100 tons, three to between 200 and 500 tons, and one of about 2300 tons;
6. Of the five major spills from ships, three were caused by grounding, one from a fire and explosion during discharging, and one from hull damage suffered before the vessel entered the port. The remaining major spill was caused by failure of equipment in a refinery ashore;
7. Approximately 80% of the pollutions came from vessels and 20% from shore installations (Dudley, 1976: 28-31).

This summary led the harbour master to conclude that with careful management in such a port, major spills should be small in number and that it should be possible to deal with them without any long-term effect on flora and fauna or the enjoyment of the general public.

Evidence from other sources suggests that current ability to manage major oil spills or chronic long-term pollution may not be as great as the harbour master indicated. Spills at Sullom Voe and other sites have caused large bird kills and their longer-term effect on flora and fauna is still uncertain. Monitoring by the British Field Studies Council at Milford Haven and other sites also indicates that the effect of major spills and chronic pollution are not well understood and tend to vary according to location and habitat (Baker, 1977; Baker and Hisock, 1980). Certain plant species are quite susceptible to oil pollution, being killed in one oiling, while others can withstand many oil applications. Cleaning methods often increase rather than decrease damage, although without cleaning, scenic and recreational uses are much more adversely affected. Changes in operational procedures following upon monitoring have resulted in some recovery of habitats subjected to chronic pollution.

Alaska

Two main agencies have been given primary responsibility for surveillance and inspection of offshore petroleum operations in Alaska: the U.S. Geological Survey, and Alaska's Department of Natural Resources (Alaska Oil and Gas Conservation Agency). Other involved agencies include the U.S. Coast Guard, the Environmental Protection Agency (EPA), and the Department of Natural Resources's Division of Lands and Forests.

The U.S. Geological Survey is responsible for supervising and regulating exploration, development, and production activities on offshore OCS leases after the leases are issued. This includes ensuring that wells are drilled in an acceptable manner, and that environmental stipulations and operating orders are being complied with. Even though there is potential for overlap in the surveillance responsibilities of EPA (for discharges of ballast and sewage), the Coast Guard (for shipboard equipment, fuel transfer, and general seaworthiness), and the USGS (for fitness of a vessel to drill holes), informal working arrangements between these three agencies have been established to reduce overlaps. While these arrangements seem to work well now, during a period of modest activity, their effectiveness could be reduced as offshore activity in Alaska increases.

Within state lease areas, the Alaska Oil and Gas Conservation Agency (AOGCA) is the main agency responsible for surveillance and inspection work. This includes checks on well design, environmental safety, and testing. The Department of Fish and Game also issues permits relating to anadromous streams, fish and game regulations, and critical areas. A staff of five is responsible for reviewing these and conducting spot checks. North of Fairbanks, however, very little surveillance is conducted due to lack of funding and manpower. According to a habitat biologist with DFG, changes in stipulations or conditions on such permits are made on the basis of either new technological or biological information or as a result of a political decision.

Monitoring and enforcement of the environmental stipulations of the Joint Federal/State Beaufort Sea Oil and Gas Lease Sale by the Alaska Area Oil and Gas Supervisor of the USGS and by the Director of the Division of Minerals and Energy Management of the Department of Natural Resources is to be assisted by the Beaufort Sea Biological Task Force (BSBTF). This task force is to remain in existence for the operating life of the field and consists of members of the following agencies:

Federal:

- Bureau of Land Management
- United States Fish and Wildlife Service
- U.S. Geological Survey
- National Marine Fisheries Service
- Environmental Protection Agency

State:

- Department of Fish and Game
- Department of Environmental Conservation
- Department of Natural Resources

In essence, the task force is an interagency co-ordination committee which is responsible for advising the supervisor and director on

> the conduct of the biological surveys by lessees, the appropriate course of action after the surveys have been conducted and on the administration of the biological/environmental aspects of the

> above-mentioned stipulations (U.S., Department of the Interior, BLM, and State of Alaska, DNR, 1979:34).

In addition to the task force, the environmental and socio-economic studies programmes run by the Bureau of Land Management grade into monitoring activities. For example, the results of the effects of research of OCSEAP are to be used

> in establishing causal relationships between OCS-related perturbations and physiological or biological change; these results form the basis for developing discharge regulations and operating stipulations. In addition, studies are evaluating biological responses to OCS stresses for their potential usefulness as early warning indicators or monitoring aids in detecting or quantifying environmental change (U.S., Department of the Interior, Bureau of Land Management, 1979a: 151).

Studies being conducted by OCSEAP on the transport and transformation of petroleum-related contaminants in the marine environment will assist in planning the location of long-term environmental monitoring stations in the study area (U.S., Department of the Interior, Bureau of Land Management, 1979a: 150).

Offshore drilling rig, Alaska, Dan Kowalski-Lightwriting, Inc., Juneau, Alaska

Chapter V: General guides or characteristics

Introduction

The basic premise of this chapter is that effective planning, implementation, and overall management is built upon certain general guides or characteristics: research, co-ordination, economic incentives, information access, and management process monitoring. The questionnaire in Appendix 1 includes some clarification of the meaning of the more nebulous of these guides, such as management process monitoring. In general, the guides can be considered as cultural processes which provide the knowledge needed in an adaptive management system operating in changing environments.

Research

In both Scotland and Alaska, local authorities generally do not have much internal research capability. The commitment to research on land use, resources, and environmental management also varies from region to region and district to district. Most local authorities gain their information either from senior government or from commissioned studies by consultants, research institutes, and universities. This research is usually short-term and applied, although often of high quality. Interest in research as an essential aspect of management seems greater at the senior than at the local government level. Indeed, one of the prime functions of senior government has been to undertake, promote, and support research.

Scotland

Local government

The Shetland Islands Council is unusual in its commissioning of a range of research projects including a site selection for an oil terminal, a master plan, and an environmental impact assessment of various aspects of oil development. The decision to undertake this research was made in the early 1970s in anticipation of development pressure following oil and gas discoveries in the North Sea. This and subsequent research formed the basis for: (1) a Shetland Islands Council *Interim Development Plan* (March 1973); (2) an *Overall Development Plan* (April 1974); (3) the Sullom Voe district plan for the site of a single oil and gas terminal and related facilities (November 1974); and (4) the *Shetland Islands Structure Plan* (December 1978). Since 1978, other local plans have been completed.

The foregoing research also provided the motivation and basis for the Shetland Islands Council to apply for and secure passage of the Zetland County Council Act in April 1974, as provided for by the Private Legislation Procedure (Scotland) Act 1936. The Zetland County Council Act 1974 contains many clauses which have made it possible for the Shetland Islands Council to take a very strong role in development, and in resource and environmental management generally. In this context, it is important to note that there was a three-year gap between the first major oil discoveries in the East Shetland Basin of the North Sea, and the beginning of construction of a single oil terminal at Sullom Voe. Much of this period was taken up with research and appraisals, the preparation of plans, and the passage of the Zetland County Council Act 1974. No major planning applications were entertained by the Shetland Islands Council until the end of this period.

From the standpoint of the Shetland Islands Council, the key provisions in the Zetland County Council Act 1974 are:

1. harbour jurisdiction over the Sullom Voe area where the single oil terminal complex was to be developed;
2. the right to construct and subsequently lease or sell works in harbour areas and to grant licences for others to construct works;
3. the right to construct, operate, sell, or lease works in any part of the islands' coastal area, subject to approval by the Secretary of State for Scotland (this and the previous clause basically give the Shetland Islands Council control of development between the low-water mark and the three-mile limit);
4. the right of acquisition by compulsory purchase of land for industrial purposes, subject to confirmation by the Secretary of State for Scotland (this power expired on 31 December 1978);
5. extensive powers to charge users for the facilities provided by the council;
6. power to establish a reserve fund out of revenue from harbour undertakings; and
7. power to invest in oil-related developments on the islands. The land acquisition power and the right to invest in development, through the Sullom Voe Association, have given the Shetland Islands Council the influence of a landowner, entrepreneur, and investor, as well as a government agency.

Senior government

Environmental and resource management agencies at the senior government level — such as the Nature Conservancy Council, the Department of Agriculture and Fisheries for Scotland, and the Scottish Development Department — have all assigned personnel to oil- and gas-related studies and provided research funding to consulting, university, and other groups. Various research records have been completed, including the *Compendium of Ecological Studies of the Scottish Coast Relevant to Development Planning* (Goodier and Whyte, 1976).

Examination of such documents suggests several things. First, much of the listed research was already underway in the late 1960s and appears to be more or less independent of oil development pressures, although it is now judged useful in response planning. Second, certain individuals and groups have responded to oil development research needs and opportunities by providing the survey type data often of major interest to planners and managers, while at the same time undertaking associated studies aimed at a more comprehensive understanding of system processes. Third, there have been different responses to the role of research in planning and management, even in areas subject to intense offshore oil development pressures. The Shetlands and the Orkneys have undertaken vigorous organized research programmes, while authorities in the Moray Firth area and elsewhere have not done so to the same degree. The effects of this difference are not altogether clear, although research programmes have helped the Shetlands to formulate management objectives and to control development pressures to a greater extent than in Moray Firth.

The results of senior government research assistance are reflected in such management documents as the Nature Conservancy Council's *A Compendium of Ecological Information for Oil Spill Contingency Planning, Shetland* (Gammack and Richardson, 1980). This compendium presents map and other data on the distribution and character of coastal nature reserves, SSSI and other special sites, as well as the distribution and relative importance of breeding, wintering, and other habitats of seal, seabird, waterfowl, and other fauna and flora. The National Conservancy Council has also worked with the Sullom Voe Harbour Authority and other bodies in providing information on a detailed oil spill response plan, and is assisting with new problems, such as how and where to dispose of waste recovered after spills.

SOTEAG

Although SOTEAG's stress is on monitoring, the group has commissioned other research (Dunnet, 1980; Foxton, 1979). For example, early inventory work showed that the upper Sullom Voe basin was anoxic at depths below about thirty metres under certain seasonal meteorological conditions. As sewage, oil spill, and other effects could be expected to have unusually severe effects in this area, detailed studies were carried out over a three-year period. The results are thought to provide an adequate understanding of this system so that the special research has now been terminated, although the area is still included in the regular monitoring activities.

An expanded monitoring and research programme has also been undertaken on birds, largely because of the adverse effects of the 1979 Esso Bernicia bunker oil spill in the voe. Baseline information has been collected under a range of projects in which the Nature Conservancy Council, and the Royal Society for the Protection of Birds have played important roles. The present research includes projects on cliff-resting seabirds, birds at sea, and other special areas of concern. A resident ornithologist has been appointed by SOTEAG for the Shetlands for an initial three-year period.

Industry

The major companies often have research units of considerable size and complexity, but resource or environmental management research appears to have a minor place in their programmes. British Gas has major research centres employing 1650 people as well as twelve small scientific units which are mainly concerned with day-to-day matters in the field. One of the five major units is developing and manufacturing equipment for inspecting pipelines without taking them out of operation. Such a system would minimize soil, drainage, and other disturbances and their adverse effects on activities such as agriculture. The intention is to develop eventually a field service for inspecting British Gas transmission pipelines.

Voluntary organizations

Many voluntary or non-statutory organizations do not see research as one of their prime functions. Previous reference was made to university and British Field Studies Council research on various technical and applied problems. A notable example of a related and somewhat different type of study is found in *Sea Use Planning,* prepared by a team of university and other researchers with the support of the Fabian Society (Young and Fricke, 1975).

This type of study can be described as a policy analysis aimed at producing recommendations for more balanced and effective management. In their introduction, the editors note that they are moving towards a quite new and unfamiliar concept — sea use planning — as a necessary function of government. They advance reasons for the need for such a concept, including the extension of the national domain into the half-million square kilometres covered by the new 200-mile offshore economic zone, and the rapid growth in offshore and coastal uses such as oil and gas drilling and production, shipping, fishing, aggregate mining, waste disposal, recreation, and defence.

Marine or sea use policy, research, safety, inspection and surveillance, monitoring, and other management measures are reviewed in the study and found to be inadequate for the task. The authors conclude that a total change has occurred in the nature and scale of Britain's use of surrounding seas. They suggest the present system of management, by way of consensus of a large number of standing and *ad hoc* committees, departments, and ministries, needs to be replaced by a new and more comprehensive approach. In particular, the authors recommend:

1. the appointment of a Secretary of State for Marine Affairs, with specific political responsibility for the general conduct of offshore government and for the development and practice of offshore planning;

2. the recognition and acceptance, in all offshore policy-making, of the organic unity of the seas;
3. the coordination and orchestration of a marine research strategy and a training program geared to foreseeable maritime requirements;
4. the development of the technique of the sea use structure map, and of a Sea Use Planning Authority, coupled with a system of public inquiry and a specialist directorate;
5. the codification and extension of British domestic law as it relates to offshore activity in general and to fish farming in particular. As often as possible, this should be in concert with neighbouring states, with the aim of establishing a regional legal framework;
6. the continued promotion of improvement in the international law of the sea, to keep it abreast of technological and political developments and of our understanding of the ocean and seas;
7. the setting up of a civilian offshore constabulary (perhaps in regionally cooperative form) to perform the various regulatory and policing roles of the civil power at sea, and the promotion of a regional, civilian sea managements regime in the countries of North West Europe.... (Young and Fricke, 1975).

Alaska

Most of the North Slope Borough research has been done by consultants who have, therefore, been important contributors to policy development and administration. In addition, consultants "have helped devise strategies for dealing with issues of land management and resource protection and have lobbied intensively in Washington, D.C. on legislation affecting the National Petroleum Reserve and coastal zone management issues" (Morehouse and Leask, 1978: 78-79).

In 1977, the borough's planning department commissioned a study of the extent and limits of the borough's regulatory authority over the use of federal, state, village, and other lands in the region (Morehouse and Leask, 1978: 100). Prior to this the planning department began to document traditional-historical use of the Beaufort coastal zone land which might be affected by offshore petroleum operations, with a view to eventual designation of industrial development and historical use zones.

According to the North Slope Borough statement of environmental protection policy (Neakok, 1977), the planning department would develop a computerized land use management information system to include oral and documented historical use, animal census, and other information. In addition, it has been proposed that in collaboration with the Inuvialuit of the Canadian western Arctic, data would be collected on herd size, location, age, and general health as an aid to more effective caribou management. Finally, an arctic environmental research management agenda was to be organized to include, among other things: (1) minimum criteria for acceptable environmental impact statements; (2) on-going analysis of environmental safety of arctic energy industrial technology; (3) on-going research and surveillance of the developing law of the sea of the Arctic; and (4) game biology and management research in which modern scientific method is joined with the memory of Inupiat oral history.

At the state level, a number of agencies are involved in research endeavours, which mainly result in survey- and inventory-type reports. Socio-economic studies are conducted or sponsored by the Department of Community and Regional Affairs, generally with the aim of assisting local governments in planning for offshore oil and gas developments, or assisting with the development of coastal management plans (State of Alaska, DCRA, Division of Community Planning, 1978; David M. Dornbusch, 1976). The Department of Environmental Conservation has prepared reports on coastal ecosystems and processes, and commissioned a study on Alaskan coastal community profiles. The Department of Natural Resources conducts geological studies and forecasts offshore development impacts in various areas. Many of these reports have been summarized in a bibliography prepared in 1979 by the University of Alaska's Arctic Environmental Information and Data Center.

The federal government conducts by far the most extensive research programmes, the two largest programmes in Alaska being SESP and OCSEAP. The Biological Services Program of DOI's Fish and Wildlife Service was established "to supply scientific information and methodologies on key environmental issues that impact fish and wildlife resources and their supporting ecosystems." The information developed "is intended for use in the planning and decision making process to prevent or minimize the impact of development on fish and wildlife." A number of projects have been initiated, including the National Wetland Inventory (Shanks, 1978).

Under the National Ocean Pollution Planning Act of 1978, the executive branch is required to establish a comprehensive five-year plan for federal ocean pollution research, development, and monitoring programmes and also to provide for the planning, co-ordination, and dissemination of information on these programmes. The act requires that existing federal programmes be inventoried, that national needs and problems be assessed and ordered, and that programmes be evaluated to determine the extent to which they meet priorities. An Interagency Committee on Ocean Pollution Research, Development and Monitoring has been established with representatives from thirteen major agencies, including the departments of Commerce, Defense, Energy, and Transportation, and the Environmental Protection Agency.

More than one thousand projects have been identified for the 1979-83 period, with budgets of $164,736,000 in 1978, $161,785,000 in 1979, and $188,388,000 for 1980. Of the agencies noted above, Commerce is the largest-funded department, with its lead agency NOAA receiving a large percentage of the total Federal Ocean Pollution Research Program budget (U.S., Department of Commerce, NOAA, Interagency Committee on Ocean Pollution Research, Development and Monitoring, 1979: S-7).

Although detailed information on industry's research efforts in Alaska or the U.S. in general has not been obtained, it is known that the Atlantic Richfield Company (ARCO) commissioned a comprehensive baseline study of the North Slope which included wildlife inventories and evaluation of the effects of oil field development on wildlife and the overall environment (Gavin, 1978). In addition, Exxon has published *The Fate of Petroleum in the Marine Environment* (Wheeler, 1978).

Co-ordination

Scotland

Several types of governmental co-ordination are necessary: (1) among local governments and their agencies; (2) between local and senior government; and (3) among senior government departments and agencies. Co-ordination can be most complex in a two-tier local government arrangement which involves both district and regional councils. Co-ordination among the nine Scottish regions and the various senior government agencies theoretically occurs principally through the Scottish Office. Municipalities and Scottish Office departments must, however, also deal with certain senior departments which are housed in London — for example, the Department of Energy.

Guidelines and other forms of policy and technical advice are issued by the Scottish Office to assist regions and districts in dealing with oil development and related technical and other aspects which are often new to them. A Standing Conference on North Sea Oil composed of local planning authorities, development authorities, harbour authorities, chambers of commerce, and oil companies also meets twice a year under the chairmanship of a Scottish Office minister to exchange views and information.

Within the Scottish Office itself, an inter-departmental committee of senior officials known as the Task Force has been established to consider all aspects of oil development affecting Scotland. The committee is chaired by the secretary to the Scottish Economic Planning Department and includes representatives of both London and Edinburgh agencies, including the Department of the Environment, the Department of Energy, and the Department of Employment.

Co-ordination arrangements between industry and municipal governments are not well-documented but few formal long-term mechanisms seem to exist. Most consultation seems to be informal and to occur between company environmental control or public liaison officers and key people in local, regional, and senior governments. In several cases, local environmental liaison committees have been established to exchange views and information. Membership can include not only industrial and local government people, but others from senior government, voluntary or non-statutory organizations, and key individuals. Such a committee has worked on the establishment of an oil refinery near St. Fergus.

Under the Town and Country Planning (Scotland) Act 1972 and other legislation, local planning authorities are required to inform and consult the public during the preparation of structure and local plans (G.B., Central

Office of Information, 1979). The public is also to be consulted in considering individual applications for planning permission, although there are certain limitations in terms of type of project, extent of effects, and the like. Registers of all planning applications are reportedly available for public inspection, and application for certain development types — for example, those involving noise hazard — must be advertized locally to give the general public a chance to comment.

Local planning authorities are now attempting to improve public participation or consultation methods; attempts are being made to reach what is referred to as the wider public, as well as voluntary organizations or other interest groups. Radio, television, community workers, street groups, and local information points are all used in various areas in preparing local plans. Conferences, seminars, public meetings, discussion groups, tape and slide shows, and specialist or generalist panels are also employed, as well as joint parties consisting of councillors, local residents, and planners. Random survey methods have been used in opinion surveys.

An array of non-statutory or voluntary bodies is involved in North Sea oil and gas development and its impacts on the shore and elsewhere. Some of these are small local groups which focus upon a particular issue or issues. Others are large national and sometimes international organizations. All of these groups engage in much informal consultation with local as well as senior government and industry.

Two major committees have been established to co-ordinate activities. The first — the Oil Development Council for Scotland — was established in 1973 to advise the government on all aspects of oil development affecting Scotland. It has twenty-seven members from industry, fishing, and other sectors, although apparently no specific representatives from local government.

Among the more significant acts of the Oil Development Council was the creation of a Committee on the Environment which presented the April 1974 report, *North Sea Oil and the Environment,* which reportedly had the following key concerns:

1. danger of abandonment and dereliction and the need to ensure that individual developers make financing available for rehabilitation;
2. the possible need for government ownership of sites in environmentally sensitive areas; and
3. the need for improved forward planning and the giving of due weight to environmental considerations in planning applications (G.B., Scottish Office, 1974).

The second major committee — the Scottish Forum on the Environment — was created in 1973 by the Scottish Office at an early stage in North Sea oil development. The goals are to exchange information and to facilitate discussion of issues. Meetings are held several times a year. Local government officials normally do not attend forum meetings, which are chaired by the director of the Countryside Commission for Scotland. Considerable information exchange must occur during the work of these two committees;

however, they are not known to have been evaluated, so their effectiveness is uncertain. It does appear that municipal governments are not involved in their activities.

An additional means of co-ordination is the programmed dissemination of oil-related information through senior government departments such as the Scottish Development Department. This agency is generally recognized as the first point of contact for industries, local governments, and other parties interested in onshore oil and gas development.

Senior government information programmes take two basic forms. First is the publication of irregular reports as issues arise and information becomes available. *North Sea Oil and Gas Developments in Scotland: A Physical Planning Resume* is one example (G.B., Scottish Office, SDD , 1975). This report provides information on oil exploration discoveries, employment, land consumption estimates, coastal survey, development zones, and other planning matters. Another example is *North Sea Oil and Gas: Coastal Planning Guidelines* (G.B., Scottish Office, SDD, 1974). Basically, the guidelines consist of identification and mapping of preferred development zones and preferred conservation zones. These zones are delimited on a very general basis over very large areas, without any scaling or priority system. Planning authorities in conservation zones are urged to prepare conservation policies, together with plans for recreation and other compatible activities, and to warn developers that major project proposals will require special authorization involving scrutiny by the Secretary of State for Scotland and "probably a public inquiry." Any intrusion into such areas is said to require compelling arguments, including a demonstration that no suitable sites exist outside a preferred conservation zone.

The coastal guidelines are generally considered useful management tools, although they are criticized as not being specific enough. Platforms and other oil and gas facilities reportedly have been permitted in preferred conservation zones by both local and senior government authorities without public inquiry or careful consideration of alternative sites.

More recently, senior government departments have placed greater emphasis on the second class of information — regular structured reports. The Scottish Development Department publishes a "contacts reference" with regular amendments on changing responsibilities in the various oil-development-related divisions of the department. The Scottish Development Department also regularly issues a National Planning Series intended "to assist planning authorities, other agencies, and developers in understanding the national view on development issues," and to provide planning guidance "on a selection of those issues on which planning authorities, other agencies, and developers might feel in need of guidance on where the national interest lies."

Three types of publication comprise the National Planning Series. Land Use Summary Information Sheets constitute the first type and deal with land characteristics or activity patterns which may be considered nationally significant. Here past changes, present situations, and where appropriate, indications for the future are summarized on the basis of available information or new studies. An example of a Land Use Summary Information Sheet is *Oil, Gas and Petrochemicals* (G.B., Scottish Office, SDD, 1977d).

If more specific guidance and information seems appropriate, National Planning Guidelines, the second type of publication, are issued — for example, the one for industrial sites and rural conservation (G.B., Scottish Office, SDD, 1977b). The summaries and guidelines are prepared in looseleaf form and amended as further work and changes occur. According to the Scottish Development Department, the guidelines are not intended to override existing local plans nor to prejudice local or senior government decisions; rather they are intended to indicate matters of national interest which the Secretary of State for Scotland "will later take into account."

The third type of publication in the National Planning Series is the Planning Information Note. Such notes are intended to supplement the National Planning Guidelines and the Land Use Summary Information Sheets. The Planning Information Notes (G.B., Scottish Office, SDD, 1977e-g) deal with administrative responsibilities of various government agencies and provide information on industrial plants and processes related to planning and management. Many Planning Information Notes have been issued on oil, gas, and petrochemicals development and effects, including the following: agriculture; pollution; freshwater pollution; marine pollution; air pollution; the Industrial Pollution Inspectorate; water, sewage, and waste disposal; health and safety; industrial development aspects; land use planning; pipelines; ports; and the Scottish Forum on the Environment.

To illustrate content, the note on marine pollution contains an introduction indicating which government bodies are of special interest — for example, the river purification authorities. The next section outlines concerns to be generally addressed and the agents responsible. Subsequent sections deal with administrative arrangements and responsibilities, pertinent legislation, contacts in the various agencies, licensing requirements, and key references.

Interviews and discussions indicate that local government personnel and others consider the National Planning Series to be a useful tool for education and co-ordination. One problem is the difficulty in clearly understanding how the various acts and administrative responsibilities actually work. Very general wording is sometimes used in the notes and this is costly in terms of accuracy.

SOTEAG is an outstanding co-ordination device. Its membership includes representatives of the major interest groups or affected parties. It therefore serves as a forum for the exchange of views among senior and local government departments, industry, voluntary or non-statutory organizations, and the general public, which is reached through SOTEAG's publications, reports, and press releases.

SOTEAG also performs certain other critical functions. It is a fulcrum for the exchange of new knowledge and for designing research by various agencies. In this context, it is important to note that its monitoring programmes are co-ordinated with those being conducted by industry and government. The industrial programmes are chiefly those carried out by the Sullom Voe terminal operator to comply with planning permission conditions and work licences.

SOTEAG has a potentially strong influence on environmental

management through its emerging advisory and educational function. The group holds meetings in which problems requiring the attention of a range of agencies are identified. A good example is the ballast water disposal problem.

Some of the ways in which industry co-ordinates its activities with senior and local government agencies and the public have been noted earlier. Industry also co-operates with other developers, notably in the management of large oil fields and installations with many owners and users. However, much of this co-ordination is encouraged or required by zoning, development conditions, or other local and senior government regulatory measures. Without such regulation and other forms of management, both private and public developers tend to handle certain activities in accordance with an opportunistic strategic planning approach.

Establishment of new oil facilities at greenfield or other sites can result in significantly lower development and operational costs and bring advantages over competitors, particularly in the early stages of a major oil or other industrial development. A good example of this, of its speculative and uncertain nature, and of the need for senior and local government guidelines and other control measures is the widespread early demand for land and other support to serve harbour and service centres for the growing number of offshore rigs.

Local and regional planning authorities and communities can also take an opportunistic and competitive stance, as the east coast offshore servicing centre situation demonstrates. Some communities have competed with one another for perceived economic opportunities, although not to the satisfaction of all members of their constituencies. Thus the fishermen in Aberdeen and other centres objected to interference from the oil company wharf and land pressures. Some moved their operations north to Peterhead where stricter management reportedly has been exerted over oil development activities through control of land sales, zoning, and other means.

No co-ordinating committee or agency for environmentally oriented voluntary organizations has been established in Scotland. Nor indeed has any special new organization been established for the coastal zone as has occurred in the U.S. in response to pressures and problems like those addressed in this report. The Scottish Forum on the Environment reportedly performs an informal co-ordinating role for the voluntary organizations, but its meetings are also attended by members of senior government departments. A variety of informal links appear to exist among the voluntary or non-statutory agencies. Conferences on monitoring or other topics are held from time to time. As in Canada, a relatively small number of persons seem to carry the burdens of the voluntary or third-party interests.

Alaska

Recognition of the many agencies with direct and indirect responsibilities for different aspects of OCS developments has provided a major impetus to development of a variety of co-ordination mechanisms in Alaska, only some of which can be discussed here.

The U.S. Coastal Zone Management Program is of course a very important mechanism for co-ordination between federal, state, and local governments. The Alaska Coastal Policy Council, consisting of elected members from each coastal district and of representatives from major state departments, provides a forum for co-operative policy and programme development. In addition, once a district has developed a coastal programme, and it has been approved by the council, all state and federal actions must be consistent with the local programme. This is the consistency provision of the Coastal Zone Management Act discussed earlier in this report.

At the state level, the need for a more co-ordinated permit review process was explicitly recognized with the passing of the Environmental Procedures Coordination Act of 1977. Under this act, a strategy for permit reform was introduced; it includes the establishment of three permit information centres operated by the Department of Environmental Conservation (DEC). These centres assist developers in identifying federal, state, and local agencies which might have jurisdiction over a proposed project. Upon request, a centre will host a pre-application conference for the developer and representatives of agencies which have jurisdiction over a proposed project. A master application process utilizing joint forms for separate permits has been instituted. To date, these centres reportedly have had a "significant effect in expediting permit reviews and alleviating misunderstandings...between permit applicants and the state" (State of Alaska, Office of the Governor, DPDP, Office of Coastal Management, 1980b: 54).

In addition, the Department of Environmental Conservation and the Department of Commerce and Economic Development (DCED) have published *Directory of Permits* (State of Alaska, DCED, Division of Economic Enterprise, and State of Alaska, DEC, 1979). The directory is a "descriptive inventory of the federal, state and local permits which may be required for the development and operation of projects in Alaska" (State of Alaska, Office of the Governor, DPDP, Office of Coastal Management, 1980b: 54).

The state also operates a State-Federal Clearinghouse within the Division of Policy Development and Planning of the Office of the Governor. This clearinghouse is responsible for the evaluation, review, and co-ordination of federal and federally assisted programmes. The effectiveness of some of these arrangements is not entirely clear, however, as evaluative studies do not appear to have been undertaken.

The Bureau of Land Management has implemented the Intergovernmental Planning Program for OCS Oil and Gas Leasing, Transportation and Related Facilities. The programme is to provide:

1. a formal coordination and planning mechanism for three major OCS program elements administered by BLM — the leasing process, the Environmental Studies Program, and Transportation Planning; and

2. the mechanism for coordinating BLM field level responsibilities with state and local governments, industry and other federal agencies (U.S., Department of the Interior, Bureau of Land Management, 1979b: 4).

In each of the six U.S. leasing areas (North Atlantic, Mid-Atlantic, South Atlantic, Gulf of Mexico, Pacific, and Alaska) a regional technical working group committee (RTWGC), has been established, along with a state technical working group subcommittee where a marketable discovery has been made. Both groups are composed of representatives of participating states in each OCS region, the Bureau of Land Management, the United States Fish and Wildlife Service, the Coast Guard, the U.S. Geological Survey, the Environmental Protection Agency, the National Oceanic and Atmospheric Administration, industry, and other special and private interests. Through the accumulation and evaluation of information, the RTWGCs provide recommendations and information to BLM, other agencies within DOI, and to state and local governments, at various decision points throughout the leasing process. The intention is that the state subcommittees should use the information from the environmental studies programme to begin a regional transportation management plan, and generally serve as a forum for information exchange on site-specific planning activities, the establishment of transportation corridors, and the location of associated onshore facilities.

The six RTWGCs form the Technical Committee of the Outer Continental Shelf Advisory Board. Two other bodies associated with the OCS Advisory Board are the Policy Committee and the Scientific Committee. These committees provide guidance and advice to the Secretary of the Interior on discretionary functions relating to OCS activities (U.S., DOI, USGS, Conservation Division, 1980: 2-12).

In Alaska, some interest in international co-operation and co-ordination has been expressed. A leader in this regard has been the North Slope Borough which sees co-operation as the key to environmental protection:

> We believe this co-operation must be circumpolar in scope and character, and it must be led by the five arctic coastal nations. Our most immediate need for this level of co-operation relates to Beaufort OCS operations begun by the oil and gas industry in the Canadian Beaufort. We look for the organization of a single arctic offshore working agreement between Canada and the U.S.; a single trilateral coastal zone management system; trilateral, because there is need for a strong arctic energy partnership between the United States, Canada, and the circumpolar Inupiat community (Neakok, 1977).

To this end, the North Slope Borough organized the first Inuit Circumpolar Conference (ICC) in 1977. The second such conference followed in June

1980. These conferences included delegates from Alaska, Canada, and Greenland. The ICC is viewed by the borough as a coastal zone management (CZM) activity:

> ...aimed at organizing international CZM co-operation able to deal with the Beaufort Sea as a single ecological system in which all offshore operations would be held to a single set of rules (Neakok, 1977

Economic incentives

Scotland

Local governments generally have not provided financial incentives for the conservation of areas of special importance for wildlife or other environmental purposes. Nor has senior government provided substantial assistance in this regard. Indeed, the general disposition of government is not to purchase land for such purposes, but rather to safeguard it by measures which influence its use under private ownership.

Some funding has been provided by senior government bodies for assessments of the impacts of oil and gas and related activities. An example is the Scottish Highlands and Islands Development Board which has financed assessments of the effectiveness of management programmes in protecting high-quality farmland and other significant renewable resources.

Little direct financial assistance is provided by government for the expenses of voluntary organizations participating in public inquiries or similar exercises. The long-established Scottish policy is that anyone whose property has been made the subject of a compulsory purchase order is awarded expenses, but only if the objection is successful. Any objector with a statutory right to appear at an inquiry can have expenses awarded against him if his behaviour is judged frivolous, vexatious, or unreasonable. Recommendations have been made for financial aid on grounds that some voluntary groups are working on topics of local, regional, or national interest, but senior government bodies generally have not agreed to contribute.

Frequent reference has been made to the use of grants and other financial incentives to assist in surveys, research, and other oil development programmes. Data have not been obtained on the amount and distribution of such assistance, so it is not possible to indicate any recent changes in the nature or level of research funding for resource or environmental purposes. Some financial assistance has been made available for harbour improvements, housing, pollution control facilities, and other projects. One problem in providing such assistance concerns the difficulty in ascertaining the extent to which new needs are directly related to offshore oil and gas development.

In the Shetlands, SOTEAG is a pivot point for a variety of economic incentives for better resource and environmental management. Its activities stimulate financial contributions from industry and the Shetland Islands

Council — principally through the Sullom Voe Association. Its own rather limited funds are invested in an array of private consulting firms, institutes, and universities. Funding of these agencies is often coupled with grants or contributions from other sources so that a strong steering effect may result.

Various tax write-offs and other incentives are available to industry for oil exploration and development as well as for the purchase of air pollution and other environmental control equipment, but the efficiency and effectiveness of these arrangements have not been studied in this report. Various incentives have been made available by industry to local councils and other bodies in connection with approvals for development sites, operations, and research. Such incentives are negotiable and much appears to depend upon the attitudes, skills, and leverage of the council or community in question. In areas in southern Scotland, or in Moray Firth in the northeast where unemployment and economic stagnation have been seen as major issues, less-generous financial incentives appear to have been made than in other less-petroleum-oriented areas such as the Shetlands.

Alaska

In Alaska, the main source of economic incentives has been the federal government — for example, through the Coastal Zone Management Act and the Outer Continental Shelf Lands Act. As a result of recent amendments, up to $5,000,000 per year can be provided for coastal states that qualify, to carry out their OCS responsibilities. These responsibilities include administrative, policy, operation, and managerial aspects of participating in the federal leasing programme — for instance, responding to federal inquiries and documents relating to federal leasing activities (U.S., House of Representatives, Select Committee on the Outer Continental Shelf, 1980: 78). However, no OCS State Participation Funds apparently have as yet been provided to U.S. state governments.

The main intention of the 1976 amendments to the Coastal Zone Management Act was to establish the Coastal Energy Impact Program (CEIP), through which funds are provided to coastal states experiencing adverse impacts from OCS energy activities: "These funds are intended to defray the expense of public facilities, services, and environmental costs associated with OCS activity" (U.S., House of Representatives, Select Committee on the Outer Continental Shelf, 1980: 77).

In addition, 1978 amendments to the Outer Continental Shelf Lands Act contained provisions related to the establishment and maintenance of two other special funds. These funds are intended to assist and compensate those who are negatively affected by OCS development. The first is the Offshore Oil Spill Pollution Fund which is to be maintained at a minimum of $100 million and at a maximum of $200 million. It is to be supported by a fee of up to three cents per barrel on oil obtained from the OCS, and money obtained by the fund through fines, penalties, and reimbursements from polluters.

The second is the Fishermen's Gear Contingency Fund. Accounts have been established for the North Atlantic, Mid-South Atlantic, Pacific, Alaska, and the Gulf of Mexico. Each area account in the fund is supported by a fee of a maximum of $5,000 imposed on each lessee and permittee in the area, with the total amount of the fund not to exceed $7,000,000 (U.S., House of Representatives, Select Committee on the Outer Continental Shelf, 1980: 74-77).

According to *Oil Drilling in the Beaufort Sea,* a report by the U.S. Greenpeace organization, oil companies are heavily subsidized by government:

> In the thirty-year period from 1948 to 1978, government subsidies to the oil industry totalled over 77.2 billion dollars. The normal federal corporate income tax is 46 per cent, and with reductions and write-offs, most corporations pay an effective rate of 30 to 35 per cent. Over the past three years, the average effective federal tax rate on the major oil companies earnings has averaged approximately 22 per cent. The average effective federal tax rate for two of the three ANS (Alaska North Slope) producers in 1978 was 6.6 per cent for ARCO (1.7 per cent in 1977) and 1.5 per cent for SOHIO (0.6 per cent in 1977). As of March 1980, the government was subsidizing Exxon, SOHIO and ARCO at the rate of 304 million dollars a month through its Alaska North Slope entitlements regulations (Greenpeace, 1980:12).

The Alaska Coastal Policy Council is currently considering regulations for compensating public groups and individuals for expenses incurred in participating in various aspects of the Alaska Coastal Management Program (State of Alaska, Alaska Coastal Policy Council, 1980: 21).

Information access

Scotland

Review of documents as well as interviews and discussions indicate that at senior government levels a considerable amount of information is available to other government agencies, industry, and citizens groups. Indeed, much of this information is provided for education and co-ordination purposes.

Nevertheless, there are problems of several kinds in securing certain planning or management information from senior government. Of leading importance in this regard are the many briefs, reports, and documents that are never announced or made available to other parties for various reasons. Some of this information is considered preliminary and is not intended for release, as positions and ideas may change. Whether such information should be available is a controversial question and difficult to pronounce upon.

Other more substantial documents are produced which are not available because of cost, convention, lack of concern, preference for privacy, or similar reasons. This state of affairs often applies to strategic planning documents for publically sensitive policy areas such as aggregate mining, dredging, pipelines, and oil refineries, where the expectation of substantial environmental impacts and resource conflicts is high. A number of examples appear in *Compendium of Ecological Studies of the Scottish Coast Relevant to Development Planning* (Goodier and Whyte, 1976). Here various joint senior-local government studies are listed but labelled "unpublished, results not available for circulation." Other studies are labelled "for internal use only," or in some cases nothing at all is said about availability. Sometimes reports are simply listed as unpublished. Such access problems apply to university and other third-party research, as well as to the research of government and industry. There appears, however, to be little apparent interest in freedom of information legislation in Scottish land use, resource, and environmental management circles.

To date, most of SOTEAG's reports have been available to all interested parties either in unpublished or published form. The group has developed an information dissemination programme, and recently co-sponsored its first major symposium. Certain documents do, however, have restricted internal distribution — for example, some reports on the major Esso Bernicia spill. Annual SOTEAG monitoring reports are filed in selected libraries throughout the U.K. as reference documents. These cannot be cited without the prior written agreement of SOTEAG.

It is difficult to evaluate information availability for the voluntary and non-statutory organizations. Information collected by some is unknown to others. Much of the information prepared by the research-oriented organizations appears to be published in standard ways. The quality of the information is often high. Certain university information is confidential, having been prepared on a private consulting basis for industry or government. Lack of funding prohibits the development of better information services in many of the voluntary or non-statutory organizations.

Alaska

As a result of the Freedom of Information Act of 1974, much government information is available to the public, and it appears that in the U.S., attempts are made to publish lists of available documents and reports. One major exception to this availability seems to be information submitted by lessees to the U.S. Geological Survey — for example, commercial or financial information which is often considered to be privileged or confidential. This and geological and geophysical data cannot be released without the permission of the lessee (U.S., Department of the Interior, 1980: 12-1 to 12-5). The same type of arrangement also exists between the lessee and Alaska's Department of Natural Resources.

Some officials from other agencies commented during interviews that perhaps USGS and DNR keep too much information confidential and that industry is generally not very forthcoming with information. Another perceived problem is that much currently available information has not been translated into useable form by either the public or government agencies. It is said that there is an urgent need for the development of a better information system.

Management process monitoring

Scotland

Interest in comprehensive, independent management monitoring is low in Scotland and the U.K. as a whole, although recognition of its potential value is increasing with the rising interest in environmental monitoring. SOTEAG involves some aspects of management process monitoring. It works in what appears to be a relatively unbiased manner, with most of the results available to all major interest groups. However, the research programme is approved by the Sullom Voe Association through budget allocations, and this undoubtedly places some limits on what the group is able to do. The SOTEAG study programme also excludes land use, socio-economic, and other changes which should be part of a comprehensive management monitoring programme. General understanding of the efficiency and effectiveness of oil development and its effects in the Shetlands would be enhanced if an organization comparable to SOTEAG were involved in socio-economic studies.

Alaska

In the U.S. there is much greater emphasis on management process monitoring. At an early stage in the development of OCS programmes, the U.S. president requested the Council on Environmental Quality (CEQ) to undertake a study which focused on the environmental impact of oil and gas production in the Atlantic OCS zone and the Gulf of Alaska. The following factors were examined:

1. estimates of potential oil and gas resources in the two regions;
2. potential energy needs and the environmental impacts of fuels that can be used to meet these needs;
3. basic steps in offshore oil and gas exploration and estimates of oil spill probabilities;
4. unusual physical conditions facing operations in the two regions;
5. environmental impact of operations in the ocean, on the shelf and along the coast resulting from exploration, production, and transportation of oil and gas;
6. economic, social, and environmental impacts of onshore development induced by offshore developments;
7. extent to which oil and gas exploration and production technology and practices protect the environment; and
8. effectiveness of federal regulatory and enforcement processes and the issues of government co-ordination and planning (U.S., Council on Environmental Quality, 1974: 3).

A number of recommendations were made on the basis of the foregoing CEQ evaluation. For example:

> Carefully designed baseline environmental studies should be initiated immediately in potential leasing areas and should be an essential and continuing part of OCS management. Such studies should be closely monitored and coordinated so that information can be integrated into ongoing operations and the results applied to decisions on leasing and regulating new areas (U.S., CEQ, 1974: 6).

Two subsequent assessments have been made of the nation-wide Outer Continental Shelf Environmental Studies Program. One assessment was completed by the U.S. General Accounting Office (GAO) in 1978, and one was completed by a National Research Council committee (NRC, Commission on Natural Resources, Environmental Studies Board, Committee to Evaluate Outer Continental Shelf Environmental Studies, 1978).

The Environmental Studies Program was initiated by the Department of the Interior in 1974 in response to pressure from public interest groups, states, and other federal agencies. Its purpose is to provide environmental data to support decisions by the Department of the Interior for leasing outer continental shelf tracts for oil and gas development.

Findings of these two assessments indicate that the Environmental Studies Program does not "effectively contribute to leasing decisions or to the accrual of social scientific information adequate for OCS management both onshore and offshore" and "may accomplish little in minimizing environmental damage during oil and gas exploration, development, and production in the Outer Continental Shelf" (NRC, Commission on Natural Resources, Environmental Studies Board, Committee to Evaluate Outer Continental Shelf Environmental Studies, 1978; U.S., GAO, 1979: i). More particularly some criticisms are as follows:

1. inadequate effects research, which is widely dispersed among various agencies and which is not co-ordinated with overall federal research needs;
2. over-emphasis on offshore data collection to the near exclusion of nearshore and onshore issues; and
3. lack of co-ordination and of an integrated management structure or administration of the programme.

Each of the two assessments included numerous recommendations for improving the effectiveness of the Environmental Studies Program. A revised design was released in 1978 (U.S., Department of the Interior, Bureau of Land Management, 1978). Whether this study design has had any success in improving the deficiencies noted by GAO and NRC is unknown. However, the potential impact of a management monitoring process is very apparent here.

For the Outer Continental Shelf Environmental Studies Program these evaluations may have had a negative impact on funding appropriations. The arctic biological and physical programme has been cut from $20 million in 1980 to $16 million for 1981, even though the area to be covered is four times larger. As a result, the programme remaining for 1981 is largely curtailed "to simple surveys of biological populations and physical phenomena," and "process-oriented, interdisciplinary work on ecological systems and dynamic events, both biological and physical, have had to be deleted or sharply reduced." One reaction to the reduction is that unless there is a dramatic reversal no later than the beginning of FY 82, "we envisage a rapid collapse of the usefulness and adequacy of the information collected by the environmental assessment program" (Norton and Weller, 1980: 2).

Finally, management process monitoring has been conducted in the U.S. by political bodies. In 1979, the House of Representatives established the Select Committee on the Outer Continental Shelf which has the authority "to conduct a continuing study and review of problems, programs, and policies related to the management of energy resources in the Outer Continental Shelf." During the ninth Congress, the committee held twenty-two days of hearings at which representatives from federal, state, and local government agencies, as well as industrial and public interest groups,gave views on various aspects of the Outer Continental Shelf Lands Act, the Five-year Leasing Program, and other issues. The committee concluded by suggesting some legislative changes, relating for example to various funds such as the Offshore Oil Spill Pollution Fund and the Fishermen's Gear Contingency Fund. A review of the foregoing studies supports the conclusion that in the U.S. there is a commitment to monitoring of the overall management process, and that such monitoring can result in significant changes in the management of offshore oil and gas developments and their effects.

Tuktoyaktuk, Northwest Territories, Ian K. MacNeil

Chapter VI: Managerial environments in the Shetlands, North Slope Borough, and Canadian Beaufort Sea

Modelling management strategies

In the course of the foregoing analysis of the coastal land and water management system in Scotland and Alaska, considerable information has also been given on the effects of oil-induced development on pre-existing economic and social activities, as well as on the resources and environment upon which they are dependent. A more detailed and systematic review of Scottish oil development impacts and management is not possible in this report. Clearly effects and controls have varied in different parts of the country in accordance with differences in oil project types, social and biophysical environments, local attitudes, and management strategies.

The different management strategies are of fundamental importance in making recommendations for improved management in the Beaufort Sea-Mackenzie Delta area. As noted earlier, in its research on the socio-economic effects of oil development, a University of Aberdeen study team has recognized three basic management strategies: (1) containment and insulation; (2) integration and absorption; and (3) major social and economic change (G.B., Scottish Office, SDD, Central Research Unit, n.d.). In the first strategy, stress is on minimizing the social effects of change by erecting and maintaining barriers between the development site and the local community. Examples of such barriers include concentrating the labour force in socially self-sufficient and self-contained work camps and ferrying workers to homes or rest centres after work periods of several weeks duration. In this containment and insulation strategy, local authorities view oil-induced changes as largely limited to some fixed time-period and as controllable through what might be called crisis management.

The second strategy, integration and absorption, also aims at minimizing economic and social effects of oil and gas activity, but in a different way. Local authorities attempt to ensure that development goes ahead, but in a manner that is compatible with local community values. Development is envisioned as integrating itself with the existing social and economic system, and major changes are avoided.

The third strategy is to use oil and gas development as a force for achieving major social change. Usually the goal is to change the economic base from fishing or other primary activities to manufacturing. Under this strategy, large-scale population, housing, wage, and other pressures are viewed as inevitable and attempts are made to guide these; this means special attention to health care, leisure and recreation, education, planning, and social work.

The University of Aberdeen team has applied these three strategies to several Scottish areas that have been subjected to intense oil development pressures (see Table 4). Information is presented on responses to four major types of development: (1) fabrication (oil platforms, etc.); (2) servicing (supplies to offshore rigs, etc.); (3) construction; and (4) construction-operation. In addition, information is presented on local perception of the scale and duration of development and effects, the local reaction, the policy or strategies adopted, and the outcome. Although the University of Aberdeen study does not provide information on oil development effects on resources and the environment, it is possible to estimate these in at least a gross way from the data on socio-economic changes presented in Table 4.

A more detailed, but succinct analysis of the Shetlands oil development experience is presented in the next chapter. In this chapter, a comparison is made of the similarities and differences between these northern Scottish islands, the Alaskan North Slope Borough, and the Beaufort Sea - Mackenzie Delta area. It should be noted that in terms of the University of Aberdeen model, the Shetlands have been subjected to several variations of the three basic management strategies: (1) integration and absorption for services; (2) containment and insulation for construction; and (3) integration and absorption for operations. This mixed approach has worked rather well in comparison to what we know of approaches and effects in other areas.

Similarities among the Shetlands, North Slope Borough, and Canadian Beaufort Sea

1. All are isolated areas in the northern part of the northern hemisphere: the Shetlands are located at about 60^0 North latitude and the Beaufort Sea-Mackenzie Delta and the North Slope Borough at about 70^0 North latitude. All have harsh physical environments, although that of the Beaufort Sea could generally be considered more challenging because of the generally lower temperatures and the presence of ice conditions much of the year. They are largely treeless, rocky, and marked by physically limiting factors such as permafrost in the Beaufort Sea and North Slope Borough regions, and peat, stormy seas, and strong gales in the Shetlands area. Peat deposits in the islands are extensive, often more than three-feet thick, and are inclined to flow if disturbed (Berry and Johnston, 1980).

2. These areas are relatively lightly populated: the Beaufort Sea-Mackenzie Delta area has about 8000 people; the North Slope Borough about 9000 residents in eight scattered villages and industrial and military installations; and the Shetlands about 22,000 people living in a number of scattered communities. One larger centre is found in each region: Inuvik with about 4000 residents on the southern edge of the Mackenzie Delta; Barrow

with about 3000 residents on the north Alaska coast; and Lerwick with about 7000 residents on the southeast coast of the main Shetlands island.

3. Until recently, the economy and society of each area has been largely dependent on renewable resources. Commercial fishing, sheep raising, subsistence agriculture, and a specialized wool and textile industry are keystones in the Shetlands. Commercial trapping, and subsistence hunting and fishing are important in the Beaufort Sea-Mackenzie Delta and North Slope Borough. In all cases, certain renewable resources are very significant economically and socially: herring, cod, haddock, and other fish in the Shetlands; and muskrat, fox, polar bear, beaver, and caribou in the Beaufort Sea-Mackenzie Delta and North Slope Borough, with the bowhead whale being an additional highly valued resource in the North Slope Borough. These resources are migratory over wide areas and subject to natural fluctuations in population, thereby making it difficult to distinguish natural changes in numbers from the effects of fishing, hunting, pollution, and other cultural processes.

4. In each of these areas, human occupancy has been a long one, extending for many thousands of years. Effective technologies, economies, and societies developed which were much smaller in scale and more individually and locally oriented than the complex systems which have been introduced in the last twenty years.

5. The people are largely of different racial and/or ethnic character than those of the larger political unit of which they are part. The inhabitants of the Shetlands reflect a large Norse influx centuries ago and those of the Beaufort Sea-Mackenzie Delta and North Slope Borough are frequently Inuit (Eskimo) or Indian. Considerable mixing with outside people and cultures has occurred; nevertheless, many people in the Shetlands, the Beaufort Sea area, and the North Slope Borough see themselves as different from the host group or larger political unit because of differences in language, lifestyle, and other factors, although these differences are less clear now than twenty years ago (Gronneberg, 1978).

6. Prior to oil development, renewable resource-based subsistence and commercial activities were yielding a living income, with some improvement lately, partly as a result of government social, economic, and technical support programmes. This effect was probably greatest in the Shetlands where fishing, sheep-raising, crofting, and textiles had become well-established by the mid-1960s, although the Banks Islanders and other Beaufort Sea people were doing quite well at that time also (Berger, 1977; Usher, 1970; Freeman et al., 1976).

7. All three areas tend to be seen as homelands by the relatively small number of inhabitants and as wildlands or unspoiled natural areas by large numbers of outsiders. Industrialists and other southerners want to develop the resources of the sealing areas, or caribou grounds, or of the "far haaf," — the seas and fishing grounds where Shetlanders have spent themselves for centuries (Balneaves, 1979). In addition, scientists, naturalists, conservationists, and tourists perceive these areas as significant nationally, and sometimes internationally, for birds and other animals as well as for their wild nature. The wildness in the Shetlands extends as much if not more to the

Table 4

Summary of developments and responses in case study areas

(A) FABRICATION

AREA	LOCAL PERCEPTION				LOCAL REACTION	
	INITIAL		AFTER START			
	Scale	Duration	Scale	Duration	INITIAL	AFTER START
LEWIS	Large – accepted company's forecasts	Very long – accepted company's forecasts	Small – branch factory	Very short – fabrication/ marine engineering over-subscribed sector	Very favourable– much-needed source of jobs for locals and potential expatriates	Less favourable – slower development than anticipated and greater uncertainty over longevity and size
EASTER ROSS	Large – accepted company's forecasts	Medium – accepted company's forecasts	Very large – unforeseen growth to early peak	Medium – diversification, hedge against competition	Favourable – jobs for locals and construction workers from smelter	Less favourable – infrastructural and general social pressures
WESTER ROSS	Medium – accepted company's forecasts	Long – accepted company's forecasts	Large – unforeseen growth, especially subcontractors (peak – very large)	Short – realization of uncertainty in platform industry and limited scope for complete diversification	Generally favourable – employment for locals and potential expatriates; unfavourable, amenities / way of life	Unfavourable – size of development, lack of containment of workers and poor supply of goods and services

SCALE (numbers employed): very small (10), small (10-100), medium (100-500), large (500-2000), very large (2000).

DURATION (years): very short (2), short (2-5), medium (5-15), long (15-30), very long (30).

G.B., Scottish Office, Scottish Development Department, Central Research Unit, n.d.

POLICY ADOPTED		OUTCOME			
OVERALL	SETTLEMENT STRATEGY	SCALE	DURATION	POLICY	SOCIAL CHANGE
Integration and absorption economic regeneration and improved population structure in rural areas	Dispersal within existing settlements in rural Lewis	?	?	Concentration of workers from Stornoway in area due to preference, slow build-up, and negligible in-migration	? Diminution in role of certain traditional aspects (e.g., church) and strengthening in others (e.g., Gaelic, occupational pluralism); ? social change without economic regeneration
Major Social and Economic Change – integral part of developing industrial/ urban society	Dispersal by significant expansion of selected settlements and accompanying services for basic work force; work camps for temporary additional workers	Large ?	Medium ?	Problem of relying on contract-dependent activities for industrialization and use of substantial new infrastructure	Changed social structure and attitudes; features of urbanization, industrialization and immigration; short-term problems; ? longer-term balance between costs and benefits
Containment and insulation "island" site concept, men in on-site work camp, all goods and services by sea	On-site work camp for labour force with some executive/ managerial accommodation in nearby settlement	Very Large	Short ?	Long-term impermanency: is development a construction project or an industrial activity? Uncontrolled spread of accommodation and area of leisure activities; short-term employment for locally unemployed.	? Encouraging out-migration in longer-term; ? increasing threat to local institutions

(B) SERVICING

AREA	LOCAL PERCEPTION				LOCAL REACTION	
	INITIAL		AFTER START			
	Scale	Duration	Scale	Duration	INITIAL	AFTER START
PETERHEAD (Marine Servicing)	Small – a few boats to be serviced	Short – sceptical of discovering commercial oil and gas	Medium – increase due to pipe-laying	Medium – acceptance of development phase being more prolonged than anticipated	Non-reaction	Generally unfavourable – large numbers of "outsiders" using bases and adjacent facilities
SHETLAND (Marine and air servicing)	Small – a few boats to be serviced	Very short – expansion unlikely to be successful	Medium – limited onshore facilities required in Shetland	Medium – large fields requiring long-term servicing	Favourable – welcome diversification of marine activity, unfavourable – conflict with fishermen	Uncertain – construction of bases locally disruptive, opportunities for locals appeared limited

POLICY ADOPTED		OUTCOME (?)			
OVERALL	SETTLEMENT STRATEGY	SCALE	DURATION	POLICY	SOCIAL CHANGE
Non-specific **– integration and absorption?**	General notion to concentrate in Peterhead with some small expansion of adjacent settlements	Large	Long	Inadequate timing of infrastructure provision; containment largely maintained	Conflict between incomers/locals at managerial levels – contrasting social networks; pressure on services from short-term residents; and increasingly "urban-orientated" attitudes of locals; unionization
Integration and absorption – disperse benefits of servicing to selected areas in Shetland	Dispersal in existing settlements	Large	Long	Concentration in two areas due to lack of base development elsewhere; incomers partly provided own residences	Most significant in construction phase at Lerwick ? eroding social networks — and in operational phase in Sumburgh area – ? incomer/local relationships most significant

(C) LANDING AND PROCESSING – CONSTRUCTION AND OPERATION

AREA	LOCAL PERCEPTION				LOCAL REACTION	
	INITIAL		AFTER START			
	Scale	Duration	Scale	Duration	INITIAL	AFTER START
SHETLAND (Sullom Voe) a. Construction	Medium/Large – generally accepting companies' estimates	Very short – generally no delays envisaged in construction	Large – continuing rise in numbers required to meet deadlines	Short – slight slippage in time-scale	Unfavourable – local jobs created far outweighed by disadvantages of incomers and impermanence	Unfavourable – local benefits (i.e., jobs) less than anticipated; continuing escalation of scale and duration of project
b. Operation	Medium – accepting company estimates and opposition to processing	Medium – use until fields fully exploited commercially			Generally favourable – longer-term local jobs but fears of intrusion into way of life	
PETERHEAD (St Fergus) a. Construction	Medium – generally accepting company estimates	Very short – rapid construction to meet early deadline	Large – generally accepting revised company estimates	Short/Medium – increased size and number of projects (including non-oil)	Generally non-committal – except environmental lobby	Generally non-committal but compounded by other "construction" projects – power station, offshore and onshore pipe-laying

POLICY ADOPTED		OUTCOME (?)			
OVERALL	SETTLEMENT STRATEGY	SCALE	DURATION	POLICY	SOCIAL CHANGE
Containment and insulation Restrict landing/ processing and associated infrastructure to Sullom Voe area	Concentrate purpose-built work camps in close proximity to terminal	Very large – depends on phasing	Medium ? possible further pipelines and small/ medium-scale processing facilities	Containment – work camps and all necessary facilities, some impingement on surrounding areas	Restrictions on local's access to work camp, development of long distance daily commuting; ? first stage in out-migration, and ? diminution of local social networks; immigration temporary workers; congestion
Integration and absorption minimise area and extent of social change	Concentration in development area by significant expansion of selected settlements and nodal point for provision of service	Large ?	Long?/ Very long? depends on whether processing and transhipment occurs	? Ultimate scale and ratio of locals to incomers the main determinants	? Fears of change in natural and social environment; ? integration of incomers and locals
Containment and insulation – restrict onshore construction activities to areas adjacent to sites	Concentration in work camps near to site, avoid providing permanent accommodation	Large – depends on phasing	Medium ? depends on further gas pipes and possible processing development	Largely containment – some spill-over of managerial staff to existing settlements	? Permanent construction labour force (power/gas/ pipes/processing), changing social networks, attitudes and behaviour – "urbanization"

(C) LANDING AND PROCESSING (CONTINUED)

AREA	LOCAL PERCEPTION				LOCAL REACTION	
	INITIAL		AFTER START			
	Scale	Duration	Scale	Duration	INITIAL	AFTER START
PETERHEAD (St Fergus) b. Operation	Small – gas separation and processing	Medium/ long – enough to exploit Frigg			Generally favourable – employment; opposition from local industry over job destruction	
EASTER ROSS (Nigg refinery and petro-chemicals) Operation	Large – considered number of developments would occur	Long – not necessarily dependent on North Sea feed stocks			Favourable – local jobs of greater certainty than fabrication; unfavourable – destroy "land" and "environment"	

POLICY ADOPTED		OUTCOME (?)			
OVERALL	SETTLEMENT STRATEGY	SCALE	DURATION	POLICY	SOCIAL CHANGE
? Non-specific **? Integration and absorption**	Concentration in Peterhead but some dispersal to smaller settlements	? Medium – Very uncertain	? Long – depends on further gas fields being exploited and possible processing development	? Depends on ultimate size of St Fergus and type/ size of processing plants	? Local social network fragmented, pressures of urbanization – changes in behaviour and demand for services
? As fabrication **Major Social and Economic Change**	Depends on supply/ demand from fabrication and other activities: if surplus, use it, if not, extend settlements?	None/ large	None/Very long	? Dependent on whether development occurs and if so, on scale and duration	Uncertainty about development occurring, its hazards

twisting and often inaccessible beaches and coasts and the extensive stormy seas as it does to the land (Berry and Johnston, 1980).

8. In the past 150 years these areas have been subjected to a number of invasions by outsiders exploiting a particular resource or range of resources. These invasions were usually narrow economically, and often took the "boom and bust" form. Examples in the Shetlands are the whaling and the herring industries, both of which collapsed before World War II. Examples in the Beaufort area are whaling and gold prospecting. Trapping has also undergone several economic cycles in this region. The Shetlands, the Beaufort Sea-Mackenzie Delta, and the North Slope Borough have been important for military and defence purposes, with the Shetlands housing thousands of men in air, naval, and army bases in World War II, and the Beaufort Sea coast in Canada and Alaska being the site of much early warning outpost construction activity in and since the 1950s.

9. All three areas have recently been invaded again, in this case on a very large-scale by national and international companies looking for oil and gas as well as ports and sites for associated development. In all cases the oil and gas search has been quite hazardous environmentally, and at the edge of available technology (indeed beyond it), requiring experimentation and new methods to explore for, find, and produce petroleum. The exploratory and development work has been very costly. In the Shetlands and the North Sea, large-scale discoveries have been made and port, pipeline, oil terminal, and other facilities are complete or under construction — chiefly through the work of thousands of so-called in-migrants. In the Beaufort Sea, servicing facilities have been developed, chiefly at Inuvik and Tuktoyaktuk east of the Mackenzie Delta. Many exploratory wells have been drilled, significant discoveries made, and large-scale development proposed, involving thousands of workers and a number of possible harbours, staging areas, oil terminals, refineries, and other projects. In the North Slope Borough, the main impacts from petroleum development have occurred as a result of the onshore Prudhoe Bay discoveries, with little offshore development having proceeded to date. Here also the local policy is for containment and isolation of the petroleum facilities.

10. The petroleum invasion and its effects have resulted in attempts by the people to reach appropriate agreements with oil companies and the host political unit, but with quite different results. Concern about the proper sharing of returns, desirable directions of change, and other factors has resulted in growing interest in greater local autonomy. Achievements in this regard have been high in the Shetlands where a principal fear has been over absorption into a more independent Scotland with possible consequent loss of powers secured in recent years from London by the Islands Council (Gronneberg, 1978). In the North Slope Borough, the practical extent and limits of the local government remain unclear; with most of the past eight years since incorporation having been devoted to establishing the borough as an authoritative presence in a region that has been dominated by federal and state authorities (Morehouse and Leask, 1978).

11. Politically, it is also important to note that the use of fish, seals, and other renewable resources, as well as tanker movement, shipping, drilling, and other oil developments, is occurring in an international context. In the

Shetlands and the North Sea, Norway, West Germany, and other countries, as well as the European Economic Community, are attempting to reconcile oil development and renewable resource issues. Similar circumstances will apply in the Beaufort Sea,which extends into both American and Canadian waters.

Differences among the Shetlands, North Slope Borough, and Canadian Beaufort Sea

1. In the Shetlands, the people own the land and have other advantages which facilitate making economic and other agreements with industry and senior governments. In Alaska, the 1968 discovery of oil at Prudhoe Bay provided the impetus for settling native land claims. In 1971, the Alaska Native Claims Settlement Act (ANCSA) was passed and regional and village corporations were established to hold title to native lands. The native corporations in the North Slope Borough have received title to four million of the 5.6 million acres they will receive. Although the North Slope Borough was incorporated in 1972, legal battles with the state and the oil industries continue over the borough's taxation powers. In the Canadian Beaufort Sea area the people are not viewed as owning the land and they have been negotiating with the federal government to secure ownership, protection of hunting and fishing rights, and other aspects of a so-called land claim since the early 1970s. Nothing has been achieved to this point beyond an agreement in principle between the federal government and an Inuit political and negotiating group known as the Committee for Original Peoples' Entitlement (COPE). The Government of the Northwest Territories, and particularly the Government of Yukon, which correspond approximately with the Scottish Office in the Shetlands case, have objected to provisions in the agreement in principle and its future is therefore uncertain. As a result of such delays, the Inuit and the other residents of the Beaufort Sea area have lacked the land or ownership leverage to secure investment, entreprenurial, employment, and resource management benefits similar to those obtained a few years ago by the Shetlanders. Income from oil landing fees, tanker fees, profit-sharing, and other agreed-upon sources has permitted the Shetland Islands Council to invest in oil spill planning, and to construct Sullom Voe harbour as well as industrial projects of a very high order. Nothing comparable has been possible for the people in Inuvik and Tuktoyaktuk, or other parts of the Beaufort Sea - Mackenzie Delta area.

2. Since 1886, the Shetlands have had a strong local government system — a county or islands regional council, recently strengthened by the extensive land use and other powers granted in the Zetland County Council Act 1974. More recently, a harbour authority has been established with responsibility for the Sullom Voe oil port and the entire Shetlands coastal zone out, reportedly, to the three-mile territorial limit. The North Slope Borough has responsibility for education, taxation and assessment, and planning within its boundaries. In addition, once it has an approved coastal management programme, it will have jurisdiction over the three-mile offshore area. However, the nearby Prudhoe Bay area remains under state control, which limits the borough's

taxing powers. The Beaufort Sea-Mackenzie Delta area has no comparable form of local government. Indeed, the fragmented settlements often have hamlet or unincorporated status and are dependent on the territorial and the federal governments for financing and power.

3. As a result of their ownership of land, their stronger form of local government, their greater financial and technical resources, and a very determined attitude towards the oil companies in particular, the Shetlanders have achieved considerable control over and benefit from oil and gas ventures. On the other hand, while land ownership and borough government have enabled North Slope leaders to influence and to participate in federal and state decision-making, they have not yet achieved the desired high level of control over oil development. Rather, they must respond to oil facilities and equipment as they arrive on the North Slope (Morehouse and Leask, 1978: 94).

4. The people of the Canadian Beaufort Sea area are in the midst of oil and gas development which will grow sharply in nature, scope, and magnitude during the next few years, if exploration continues into the full development and construction phase. While in this situation, these people have no ownership powers, a weak fragmented system of local government rather than a regional one, no port or coastal zone authority or agency, no significant research or technical support, and limited funding for any possible management activities. No arrangements exist at the senior government level to counter-balance these weaknesses, if such counter-balancing is ever fully possible from outside such a region. The Canadian Beaufort Sea oil development, resource, and environmental management system is similar to that previously described for the U.K. North Sea — that is, it is fragmented among numerous committees, agencies, departments, and ministries, including the departments of Indian Affairs and Northern Development, Fisheries and Oceans, the Environment, and Transport. Indeed, this fragmentation and complexity is perhaps more all-encompassing than in the North Sea because these same federal government units are involved in both offshore and onshore oil development and management, along with comparable agencies within the governments of Yukon and the Northwest Territories.

In certain respects, the territorial governments suffer difficulties similar to those described previously for the local Beaufort people. For example, they own very little land in the large area for which they have increasing political and managerial responsibility. This situation deters the territorial governments from taking strong initiatives in areas where they could do so — for example, coastal zone planning and management.

Fishing facilities and activities in Lerwick harbour, Shetland Islands, September 1980, J.G. Nelson

Chapter VII: Effectiveness of the oil development, resource, and environmental management system in the Shetlands

In this chapter the intent is to summarize succinctly the major effects of oil development in the Shetlands and the extent to which these effects have been controlled by the evolving management system. The stress is on resource and environmental effects, but some information is also provided on population, land use, and other changes which strongly influence the other effects that are of principal interest in this report. Information has been obtained through interviews and from key documents such as *Shetland in Statistics* (Shetland Islands Council, Research and Development Department, 1980).

As noted earlier, because of time and other constraints it has not been possible to undertake the same detailed analysis for the North Slope Borough in Alaska. The more stringent U.S. strategic planning, environmental impact assessment, and other requirements have prevented the incremental offshore exploration that has occurred in the Canadian Beaufort Sea. As a result of a recent U.S. Supreme Court decision, however, approval has now been given for the commencement of marine drilling programmes in the Alaskan Beaufort Sea. A study of oil planning and preparatory effects up to this time would therefore be very timely, and would establish a useful baseline for assessing future impacts, and for comparing the effectiveness of U.S. management procedures with those in northern Canada, the Shetlands, and other offshore petroleum development areas.

Population: forecasting and changes

In 1971, the Shetlands population was 17,327. By 1979 it had risen to about 21,760, with much of the increase due to oil development. This 1979 estimate does not apparently include the more than 7000 workers now housed at two construction camps near the Sullom Voe oil terminal. The distribution of population increase has been uneven but more or less in accordance with planning. The greatest increases have been at the main administrative and commercial centre of Lerwick (6407 to 7058) and at villages selected for new housing, such as Delting and Nesting (combined totals: 1054 to 2093) and Bigton Junction-South (848 to 1757).

Population and other forecasts have, however, been inaccurate for several major reasons, including: unanticipated environmental obstacles, changes in technology or market conditions, and industrial competitive strategies. In the Sullom Voe environmental impact assessment, it was estimated on the basis of company data that the peak labour requirement for the oil terminal and associated facilities would be 2200 workers. The current work force has been at the 7000-level for some months. The reasons for the large increase over the forecast are not entirely clear. One major influence has been the difficulty in meeting construction timetables because of gales, heavy precipitation, raw cold, peat problems, and other physical obstacles, the strength of which was apparently underestimated.

The forecasting inaccuracies have had various adverse environmental impacts. For example, the work camps or villages have had to be larger than anticipated. Housing is for 4500, with the remainder of the current employees being on leave at any one time. Not all the housing is at the villages, however. Two ships have been docked in the inner part of Sullom Voe to provide berths for 800 workers.

According to the original planning, the work villages were to be removed when construction was completed. The Shetland Islands Council and other parties are now considering maintaining one of the camps for workers in transit to and from existing and new rigs in the northern and western parts of the North Sea. Associated with this possible change is another to expand a World War II air force base near Sullom Voe, which has been used increasingly to move workers, because landing at the main airport at Sumburgh requires a two-hour drive across the main island to the voe area. The resource and environmental effects of these possible modifications in planning have not been estimated.

Housing and related services

Even with the containment and insulation policy for construction activities, demands for housing, school, hospital, and other services have increased markedly. Between 1971 and 1980, housing stock rose from 5787 to 8182 units. Demand still exceeds supply and house prices reportedly are twice those for comparable mainland units. Primary and secondary school enrolment has risen from 2014 and 897, respectively, to 2738 and 1619 from 1971 to 1980. Casualties at the major accident hospital have increased from 1082 in 1971 to 2799 in 1979. The amount of school, hospital, and other structural growth caused by the preceding changes is not known. While the general view appears to be that council and planners have largely been able to limit new development to desired levels and designated areas — thus curtailing sprawl, blight, loss of agricultural land, and other unwanted effects — there are some problem areas which may increase as time goes on and as further unanticipated modifications in economic and other activities take place.

Transportation facilities

The effects of oil-induced increases on the scale of operations at Sumburgh airport and other air transport facilities were not estimated in the SVEAG environmental impact assessment or other studies. The Sumburgh airdrome has grown considerably since 1972. Annual air traffic movements rose from 57,343 to 63,390 between 1978 and 1979 — an increase of about eleven per cent. Sumburgh airport is one of the busiest in the U.K. and is reportedly the largest current permanent employer in the Shetlands, with 734 employees in 1979. The very high proportion of traffic attributable to oil development can be estimated by examining passenger totals. In 1979, 622,446 passengers used the airport. Of these, 279,908 were charter and 253,214 helicopter passengers respectively. These two categories are associated largely with oil development. The growth in use and size of Sumburgh airport has undoubtedly increased noise levels significantly in its vicinity and has led to housing, road, and other forms of urban sprawl, as well as agricultural blight around its periphery. The future of Sumburgh and several smaller airports is difficult to predict. A very sharp drop in traffic seems likely with the cessation of oil terminal construction. Much of the helicopter traffic is to service the rigs and can be expected to continue. However, even here there may be a change: large longer-range helicopters are coming into service at southern centres such as Aberdeen and these could lower demand appreciably in the Shetlands.

Transportation other than by air has also increased sharply over much of the islands — particularly in the Lerwick area. External ferry traffic rose from 37,544 to 46,823 passengers between 1977 and 1979. Inter-island ferry traffic increased from 23,110 to 32,501 between 1970 and 1979. These increases are associated with rises in automobile and road traffic. For example, 7390 cars came to the islands by external ferry in 1979, as compared to 5164 in 1977. Net registered tonnage in Lerwick harbour rose from 624,233 to 1,561,122 between 1971 and 1979.

Large transportation increases were anticipated. Indeed, the shipping increases at Lerwick and some other sites are part of the integration and absorption strategy adopted by the Shetland Islands for the service aspects of offshore oil development. What the resource and environmental effects have been is difficult to ascertain, especially for outlying regions. However, demand for docks, wharf space, and other facilities has been very high in Lerwick, causing some competition between oil interests and fishermen. Considerable harbour in-filling and reclamation has occurred to create new space and this practice is continuing with unknown effects.

No data was obtained on road mileage changes between the early 1970s and 1980s. However, the major impact here appears to be due not so much to new road construction as to upgrading of the pre-existing network so that it could accommodate the larger trucks and the heavier traffic associated with oil development. Roads therefore have been paved, widened, and straightened over much of the mainland. Attempts to control adverse effects do not seem to have been intensive, and adverse effects are widespread. Cuts, and exposed slumping peat, line many of the roads. Fill and aggregate sites generally have been located close to the roads as well. These sites are often large and unsightly to many residents and visitors. Some of the traffic changes have interfered with farming operations. Sheep kills reportedly have increased significantly along the roads, although no data appears to be available on the magnitude of this effect. Peat cleared by road or other construction is often tipped or left haphazardly along the routes. The peat stripped from several hundred acres during construction of the Sullom Voe oil terminal has been deposited at the head of a small bay. Attempts are being made to grass and rehabilitate this for recreation and other uses. The handling of the peat problem and of rehabilitation has been a difficult and costly one.

Pipelines

Two pipelines have been constructed from the North Sea fields to the Sullom Voe oil terminal, which is to have a planned production of 1.5 million barrels a day when completed sometime in the 1980s. The onshore pipeline routes generally seem to have been well-designed and rehabilitated. Some heather has been converted to grass and is now being grazed on by sheep. On the other hand, there are some bad sections where gravels and subsurface materials have been left near the surface and have made regrowth slow. Drainage channels also have been changed in a few areas,and if erosion at these sites continues, it could be costly in the future.

Offshore pipeline construction has apparently not always been conducted with proper care and has caused some problems for, and conflicts with, fishermen in particular. Trenches have not been filled and boulders, heavy clay till, and other kinds of debris have been left scattered over the seabottom. Pipeline construction also reportedly has occurred frequently on the softer bottom sediments preferred by fish and therefore used by fishermen. Consequently, there has been damage to fishing gear as well as pipelines, and loss of access to some fishing grounds, which many fishermen perceive as part of their heritage. The magnitude of the economic effects on fishermen is unknown and is currently under study at the University of Aberdeen.

Fishing industry

Other impacts on fishermen include loss of access to fishing areas around the oil rigs and platforms, occasional craft or net damage due to collisions with steel frames or other debris, and increases in operating and processing costs because of oil-induced increases in general wage levels in the Shetlands (Coull, Goodlad, and Sheves, 1979). The increases do not seem to have pulled many men directly off boats, for they seem to prefer this life above all others. But the wage increases have apparently pushed up labour and other service and supply costs at the wharf and shore. The high wages offered to maids and other hotel, work camp, and service employees have led to rising costs and so to labour losses in fish processing and related activities. Whether the cost of loss of access to fishing grounds or from collisions or other hazards is too high or not is difficult to say, in part because little pre-planning appears to have been done on acceptable levels and risks, and desirable management procedures.

Inflationary and other socio-economic effects on the nature and viability of the fishing industry were anticipated by government and other interested parties, but no wage support or other possible means of compensation or adaptation were created. The inflationary effects appear to be most marked in the Lerwick area where the integration and absorption strategy is being pursued by the Shetland Islands Council. These effects are therefore something that can be expected to continue over the longer-term and require careful planning. Management of the fishing industry is complicated by factors such as increasing competitiveness among EEC countries, and the contention by fishermen that London has not developed a sound fishing industry policy.

The number of fishing vessels in the Shetlands was 136 in 1971, rose to 146 in 1974, and has dropped steadily to 110 in 1979. Employment in the industry (catching, processing, and ancillary) was 1781 in 1971, rose to 1860 in 1974 and dropped steadily to 1407 in 1979. The decreases have been greatest in the catching and processing sectors, moving from 636 to 584 and 967 to 504 respectively between 1971 and 1979. In retailing and ancillary activities an employment increase from 178 to 319 has occurred in the same period. The total fish catch has fallen from 49,634.66 tons with a value of £4,219,470 in 1976 to 28,688.25 tons and £4,071,287 (inflated pounds) in 1979 (Shetland Islands Council, Research and Development Department, 1980).

Tourism

Tourism, an industry of growing importance in the 1960s and early 1970s, has been affected adversely by oil development. Many visitors are coming to the islands, but primarily for commercial reasons. The accommodation capacity available to the Shetlands Tourist Organization by bed spaces was 1152 in 1976, rose to 1342 the following year, then fell to 1137 in 1980. The tourists lose out in the competition for beds and other places to the more well-to-do businessmen. Oil and other companies often book blocks of rooms even though these are not always used. The effects of these changes on hotel owners or others involved in tourism is not known, as no evaluation or monitoring studies have been undertaken. Presumably, many hotels have felt no loss; indeed their income may have risen with the business bookings, but for others involved, the costs and benefits are not so apparent.

Oil spills

Between the opening of the Sullom Voe oil terminal in November 1978 and 4 December 1979, eight oil spills occurred; most were minor, although the long-term cumulative impact of such spills is a concern in the anoxic waters of the upper voe. The large 1174-ton bunker C oil spill was a very serious one. All parties were unprepared to deal with the spill, in part because it involved the first large tanker to enter the port and in part because planning and control activities were inadequate. On Saturday, 30 December 1979, as the 193,000-ton Esso Bernicia was being guided to dock, a tug engine fire led to loss of control of the tanker. It struck the Number 2 jetty and subsequently released nearly 1200 tons of bunker fuel which spread during the following days over much of Sullom Voe and into Yell Sound in spite of pre-operational planning assurances to the effect that up to 2000 tons could be contained without difficulty.

Among the factors contributing to the spill were: (1) the failure of both control booms; (2) the failure of the engine of a relief boat; (3) the absence of a fourth tug because the crew had been given a holiday; (4) some of the promised sophisticated control equipment was not yet available; (5) a planned oil spill control equipment base had not yet been built; and (6) the boom designed to seal off Sullom Voe from nearby Yell Sound was not available in the manner planned for in 1976. Control of the spill was exacerbated by the bad winter weather conditions. Company and other interested parties apparently also had not anticipated and planned for this particular type of spill. Provisions had been made for a lighter cargo oil spill, rather than for the thick heavy bunker fuel oil, which requires different control measures.

Sheep, seabirds, and other animals were killed or injured, with the effect on fish not being well-known. Sheep in the Shetlands feed on shore seaweed and other vegetation, and about twenty were killed, with thousands injured — for example, by having their fleeces destroyed. Compensation was paid by Esso and fencing has since been erected around much of the voe. More than thirty otters were killed or "oiled." Seabird deaths were high. Among the known deaths were hundreds of great northern divers, long-tailed ducks, eiders, shags, guillemots, and black guillemots. In some of these cases the deaths were thought to represent significant proportions of the wintering and breeding population. All parties with whom this toll was discussed regarded it as serious, but some disagreed on its significance. All considered the spill to be a very unpleasant event, but some saw the really important question as being whether it had had any kind of fundamental effect on the population size and future of certain birds or other animals.

Tanker ballast and related problems

The release of oily ballast and other waste by tankers is a problem that was foreseen, but has been difficult to control. The tendency to dump such waste offshore prior to entering the port has been lessened by the Sullom Voe Harbour Authority requirement that entering tankers have thirty-five per cent of cargo space in ballast. Disposable ballast can then be fed into the treatment plant at the oil terminal before being emptied into the Yell Sound. The introduction of tanks to hold ballast separate from oil in the holds also has tended to reduce the pollution problem. A new concern has arisen, however, in that such ballast may be collected in industrial ports elsewhere in the world and the best methods of treating this unknown commodity in the voe or other harbours is not clear.

An associated problem is that some tankers do not travel along recommended routes, but rather skirt close to coastal hazards or pass through incompletely surveyed waters. Legal and managerial uncertainties make control of such movements difficult. Scottish Office Planning Information Notes indicate that councils and harbour authorities are responsible for spill control and clean-up as far as one mile from shore, beyond which a mix of industry, the Coastguard Service, the Department of Trade, and possibly other senior government departments such as Energy and Defence becomes involved.

Aside from the obvious difficulties of how to organize activities and responsibilities around the one-mile limit or further offshore, there is also the problem of providing sufficient patrol craft and other costly specialized equipment for effective control. Oil spill equipment poses particular problems because major spills, although potentially very hazardous, are infrequent, making it expensive and difficult to be at the ready with quickly deployable equipment at enough locations to be effective.

Coastal management and the harbour authority

In the Sullom Voe case, the harbour authority has accepted responsibility for all island coastal waters out to the three-mile limit and has made preparations to deal with oil spills, other pollution, and related problems in that zone. Inasmuch as the authority's jurisdiction over and ability to control offshore ballast disposal, tanker movement patterns, ship safety equipment, and other matters of concern is uncertain, and that enforcement is difficult, careful inspection systems have been developed which can cause long and costly delays in entering and using the port for obvious or repeated non-compliance. Table 5 shows the staffing arrangements for the Sullom Voe Harbour Authority and reveals that piloting and inspection services constitute a major part of the large roster and budget considered necessary by the Shetland Islands Council for effective coastal management. The council has also found it necessary to lease a plane regularly for offshore surveillance and control.

As a result of the Esso Bernicia oil spill and other events and issues, the council has had two basic planning and management documents prepared. The first is entitled *Shetland Islands Anti-oil Pollution Scheme* (Shetland Islands Council, n.d.). This document provides useful information on areas of responsibility, equipment storage and maintenance, personnel training, safety, use of dispersants, keeping of records, oiled bird treatment, assessment of risk of various forms of pollution, reporting and reconnaissance, relationships and liaison with other authorities and organizations, operational planning, and ecological considerations within the scope of the scheme.

The second key document is the *Sullom Voe Harbour Oil Spill Plan,* prepared by the Sullom Voe Oil Spill Advisory Committee (n.d.). The plan specifies that companies own and properly maintain oil spill equipment and materials. The location of equipment, communications systems, command responsibilities during a spill, and manpower and training, are among the matters discussed in the plan. The document also contains a map showing the

Table 5

Shetland Islands Council Ports and Harbours Committee; staffing position 29 August 1980

Position	No.	Remarks
Director	1	
Deputy Director	1	
Deputy Director/ Deputy Harbourmaster	1	
Port Engineer	1	
Oil Pollution Control Officer (O.P.C.O.)	1	
Assistant O.P.C.O.	2	
Marine Officers/Pilots	24	
Mechanical Engineer	1	
Electrical Engineer	1	
Administrative Assistant	1	
Clerical Assistants	2	
Secretary	1	
Launch Crews	30	
Junior Deck Hands	6	
Driver/General Duties	1	
Caretaker/General Duties	–	Post advertised
Labourer/Relief Boatman	2	
Cook/Cleaner	1	
Storekeeper	1	
Semi-skilled Maintenance	2	
Total	80	

Source: Ports and Harbours Department, Seallaness, Scotland.

location of sensitive areas in the oil terminal area, including fish spawning grounds, vital areas for birds and other animals, amenity beaches, salt marshes, and sites of special scientific interest. Appropriate actions are recommended for these various areas, with booms and other equipment installed in anticipation of an emergency.

Financing coastal management

The foregoing local coastal management system has required much time and effort to prepare, with serious planning underway since about 1975. The system is costly both from the standpoint of capital and operational costs. Approximately £20,000,000 has reportedly been spent on the headquarters and related harbour authority facilities. It has been possible for the council to provide these basic services because of the various provisions in the Zetland County Council Act 1974 and because of the council's control over port operations, services, and fees. The fees include: (1) pilotage charges of 0.9 pence per gross registered ton, with a minimum of £25.00 for either an inward or outward journey from the voe; and (2) boarding and landing charges of one pence per gross registered ton per act of pilotage, and moving rates — for example, £300 for up to 40,000 gross registered tons rising to £550 for craft over 110,000 gross registered tons (Sullom Voe Harbour Authority, 1980). Approximately 600 tankers per year are currently using the port. This is lower than the forecasts, and has led to an increase in the anticipated fees.

Regional and related planning pressures and needs

The coming of the oil developments has placed much strain on planning and related local and senior departments responsible for the Shetlands. Funds from oil-induced sources have been used to pay for the various consulting and monitoring services perceived as necessary by the council — for example, SVEAG and SOTEAG. Professional planning staff have also been added to assist in preparing the Shetlands structure plan and six district plans. The latter do not have the formal significance of those prepared for official district councils in two-tier local government areas elsewhere in Scotland. The current official plan took five years to prepare but urgently requires updating, according to the director of planning. Several of the community or district plans also should be updated. Others are required for some parts of the islands which have never yet had a plan. Due to difficulties in finding the resources to undertake all these tasks, the planning system is being reconsidered. It may not be feasible or desirable to continue with the local or district plans in the manner originally anticipated when the first plans were prepared.

Mackenzie Delta; Parks Canada

Chapter VIII: Summary and recommendations

Many lessons, suggestions, and recommendations have been presented in the body of this report and no attempt will be made to give a summary list here. The assessment model utilized in this report in itself provides a guide for thought about the detailed development of an effective management system. However, as a result of experience gained in this study, two additional elements should be added to this model. One is forecasting, an activity upon which many plans and decisions are made, yet which often is inaccurate. Indeed, the high propensity to error in this fundamental planning activity means that as many affected groups as possible should be involved in preparing demographic and other forecasts. The second element is design, or the fitting of physical or other structures to biophysical and social processes. Both of these elements constitute key aspects of planning, along with the establishment of strategies, inventory, classification, and evaluaton (see Table 6).

As a result of the application of the management assessment model, certain key generalizations can be made about the Alaskan and the Scottish experiences. Of major significance is the strong local response that has occurred in the Shetlands and in the North Slope Borough. In the Shetlands, this reponse has largely been within the context of a modified pre-existing local government system. In the North Slope Borough, the local response has partly been facilitated by technical and financial assistance through the Coastal Zone Management Act. In both areas the response has also been supported by local income from oil development, land ownership, and related enterprises.

Considerable anticipatory and strategic planning has taken place in the Shetlands, the North Slope Borough, and other parts of Scotland and Alaska through inventories, land and water classification, identification of especially significant areas, environmental impact and other evaluations, careful permit approval procedures, research, and other management processes. Concern about monitoring is high in both Scotland and Alaska, primarily because much uncertainty applies to many of the projects. Surveillance and inspection procedures require strengthening in both areas. In contrast, many of the aforementioned management processes are considered to be inadequate in the Canadian Beaufort — notably from the standpoint of government responsibilities.

Certain basic principles also emerge from the Scottish and Alaskan experiences, particularly from the Shetlands. In setting these forth below, no attempt has been made to be very specific about institutional or other arrangements best-suited for implementation in the Canadian Beaufort Sea region. More detailed study would be required to do this. Furthermore, the appropriateness and effectiveness of managerial systems varies from place to place in accordance with differences in society, in economy, or in the

Table 6
Management assessment model (revised)

			Local	Provincial	Federal	International	Committee	Commission	New Agency	Industry	Other
Agency	Government or Management Level or Type										
	Lead Agency										
	Participating Agency										
Planning	Strategic Planning										
	Forecasting										
	Inventory (Biophysical & Social)										
	Land Classification										
	Especially Significant Areas										
	Evaluation										
	Design										
Implementation	Approval	Negotiated Development									
		Pre-Hearing									
		Hearing									
		Formal Case Record									
		Permit Order With Conditions									
		Appeals									
		Final Decision									
	Construction and Operation	Surveillance & Inspection									
		Monitoring									
		Enforcement									
		Modification									
General Guides or Characteristics	Research										
	Co-ordination	Government									
		Industry									
		Public									
		Inter-group									
	Economic Incentives										
	Information Access										
	Management Process Monitoring										
Comments											

environment, so that caution is necessary in dealing with needs in the Beaufort on the basis of experiences elsewhere. On the other hand, enough information on the Beaufort Sea region has been obtained from available reports and interviews based on the management assessment model questionnaire (Appendices 1 and 2) to be knowledgeable of the general situation there and to be able to relate it to experiences in other areas.

Summary of basic principles

1. One fundamental principle that emerges from this report is that local people can manage large-scale oil and gas development reasonably well given appropriate attitudes, strategies, institutional arrangements, funding, and time. Having said this, it is important to recall that management in the various Scottish regions is not always proceeding in the same general direction. This was noted in the discussion of the three basic Scottish strategies for adapting to the social effects of North Sea petroleum development: containment and insulation; integration and absorption; and major social and economic change. The focus in this report has been on the first two in the Shetlands. The major social and economic change strategy has, however, been adopted in the Inverness and Cromarty Firth areas. Major industrial projects are underway there which have had destructive effects on prime agricultural land, fisheries, beaches, dunes, marshes, and other renewable resources and related activities. Two key questions in the Cromarty Firth and other areas pursuing major change are whether an appropriate balance between new industrial development and pre-existing and potential renewable resource-based industries such as agriculture and fishing has been achieved, and whether enough new industrial development can be attracted to the region to create sustained growth rather than "boom and bust."

2. The keys to the Shetlands — and to a lesser extent to the Alaskan — success in coping with oil development have been: (1) a high level of self-confidence; (2) a strong bargaining attitude; (3) good information; (4) an appropriate set of strategies and institutional arrangements, with new needs being supported by senior government; and (5) sources of funding to do the necessary planning and management in a short but not too constrained period of time. Not too much has been said to this point about the Shetlanders' self-reliant attitude, but anyone who reads about and visits their islands quickly discovers that it is there. One sees it in the crofters and farmers, and in the fishermen and people who have made their living for generations in this beautiful but rather hostile place. They are proud of, and generally seem to want to preserve, indigenous things: the language; the Norse connection; artifacts of the past; and the traditional landscape scene. It is most important, however, to realize that this self-reliance depends upon the people having substantial control over land use, resources, and the environment through land ownership and in particular a strong local government.

3. Senior government support has been important, but in ways that are perhaps different than most people would expect. In the Shetlands, the support has come not so much through large-scale senior government activities in the islands, although technical work and funding from the Nature Conservancy Council and other agencies is important to planning and management. Rather, principal support has come through the facilitation of measures seen as needed by the Shetlanders for comprehensive planning and management and also to work with the powerful oil companies. Of special significance here is the Zetland County Council Act 1974, which provides for the land use zoning, land purchase, harbour authority, coastal zone management, fund raising, and other powers basic to the development of an effective response. Of comparable significance in the North Slope Borough is the assistance available under the Coastal Zone Management Act.

4. A central part of the Shetlands response has been the creation of an evolving economic, social, and environmental plan, and more specifically, of goals, policies, and programmes for the management of the coastal zone where large-scale, oil-induced effects were to be expected. And we have seen the measures taken in this context: (1) one oil terminal and related facilities to be shared by all companies and to be constructed and managed through the containment and insulation policy; (2) the creation of a large, eighty-man harbour authority to manage the oil terminal area, the Sullom Voe, and the Shetlands shore on behalf of the council through funds generated largely by oil development; (3) the use of inventories, sensitive areas classification evaluation, and design systems; and (4) the development of co-ordinating mechanisms such as SOTEAG, as well as the Sullom Voe Association, for research, monitoring, and other purposes. The successful use of such institutional arrangements is of course very closely related to the decision to develop oil terminal, oil port, and related facilities in only one area; this reduces the tendency to spread activities and effects, making inspection, monitoring, enforcement, and other implementation measures easier, more effective, and less costly.

5. As a result of the funding, research, co-ordination, and other arrangements noted above, the Shetlanders and native Alaskans are now part of an overall intelligence system which seems essential to the efficient and effective management of oil-induced and other large-scale industrial effects. This system ideally permits continuous reaction to changes, although not without uncertainty and difficulty. However, the uncertainty about the nature, extent, and implications of change, as well as technical or other means of dealing with it, is the basic reason for needing such intelligence in the first place. In this context, the Shetlanders have sponsored relevant research, monitoring, and conferences through SOTEAG, and participate in the activities of the Scottish Forum on the Environment, and other organizations in which information on issues of local, regional, national, and international relevance is presented, discussed, assessed, and put to use.

Recommendations

On the basis of the foregoing summary, and recalling how strong local self-government, land ownership, and other factors have supported a self-reliant, intelligent, and effective response to oil development, the following principles should be implemented in the Beaufort Sea area:

1. Strengthen and improve local government

Local government should be strengthened, preferably by the creation of a regional structure of some type (Rees, 1978), linking Inuvik, Tuktoyaktuk, and other settlements in the western Arctic. This regional system should have a central role in managing adaptation of any oil and gas development by arranging for planning strategies, and implementing responses in association with the local communities, senior government, and industry. The implementation of a regional government will likely require some time to implement because of the need to build many bridges among the communities and senior government agencies.

2. Improving the permit or approval system

A comprehensive permit or approval system similar to that used in the Shetlands and the rest of Scotland should be created at the regional government level with input from senior government agencies and other concerned parties. Certain federal or territorial agencies would, of course, retain authority for special permits — for example, water use permits under the Northern Inland Waters Act.

3. Allocate land to the native people and to territorial and local governments

A stronger role in development and environmental management should be encouraged by allocating large amounts of land to native people as well as to territorial and local (regional) governments. The best method of allocation is, and may always be, uncertain. But some prompt allocation has immediate advantages that need not be lost through the desire for a comprehensive once-and-for-all settlement of the type that seems to be sought currently in the Yukon and the Northwest Territories. Hunt (1978) has presented ideas on possible approaches to this problem in her comparative studies of recent land claim settlements, particularly in Australia. In assigning such lands to indigenous people and local (regional) and territorial governments, the effect would be to create a counterweight against the broad-brush national approach to oil production and other massive development taken currently by industry and certain federal government agencies. Some of these agencies are also involved in deciding whether or not the new projects are in the economic, social, resource, and environmental interests of the local people. This essentially political role is largely undertaken by area residents in the Shetlands, as it is in most other parts of Canada outside the North. Such a transfer of land would also provide the local people and their governments with political leverage as well as some of the funds needed to negotiate more effectively with the oil companies in planning and implementing development, as well as resource and environmental management policies, programmes, and projects in the style of the Shetlands. With such a land transfer, the focus of interest would pass more rapidly from employment issues to entrepreneurship, investment, and stewardship.

4. Establish a Beaufort Sea-Mackenzie Delta shore zone planning commission

A shore zone planning commission or group should be established to co-ordinate planning of oil exploration, drilling, dredging, island building, construction, servicing, shipping, mining, hunting, trapping, fishing, defence, scientific, and other activities in the onshore and nearshore zone of the Beaufort Sea-Mackenzie Delta area. The commission would also be responsible for tracking and reviewing petroleum development and other activities far offshore in the deeper waters of the arctic shelf. In tracking and reviewing such activities, the Beaufort Sea-Mackenzie Delta shore zone planning commission would mainly be interested in the effects that these activities would have in the onshore and nearshore areas. What, for example, would be the effects of developing a large deep-water oil and gas field in terms of new docks, warehousing, navigation, in-filling, dredging, and other onshore facilities and activities?

The planning commission would serve as the lead agency in planning for such wide-ranging activities in the Beaufort Sea shore zone, activities whose growth and effects are frequently said to be subject to insufficiently co-ordinated permit and approval procedures under an array of different agencies, mainly at the federal level. Many approvals are reportedly given without adequate understanding and careful consideration of the water movements, erosion and deposition patterns, and wildlife habitat, native use, scientific value, or other aspects of the shore zone. Appropriate biophysical and social inventories, resource classifications, and evaluation and design studies are needed to provide the basis for an improved approval process. Such studies are also needed as a basis for considering and deciding upon alternative port, oil terminal, oil refinery, pipeline, and other projects. Nor are approvals based currently on any overall strategy for the guidance of development as in the Shetlands, and the Sullom Voe oil port, for example. The approval procedure also does not currently involve any significant roles for the local people and local government other than advisory and consultation functions.

As for example in the Alaska Coastal Policy Council, the local people and local government should be prominently involved in the membership and work of the commission so that one of its benefits would be assisting in the creation of strong regional and local government. In this context, within a few years certain activities of the commission would likely become redundant in that they would be taken up by regional government, although the commission could continue to play a useful research and monitoring role for many years. To encourage such an evolution in role, the urgent planning tasks facing a commission should be assigned to a specified time-period.

A Beaufort Sea-Mackenzie Delta shore zone planning commission should immediately undertake the following basic tasks: (1) a comprehensive review of all available information on oil-induced development, major land and marine use proposals such as mining and shipping, and resource use and environmental protection in the shore zone; (2) identification of major information and research needs — for example, in strategic planning, inventory, sensitive areas classification, evaluation, and other aspects of the management assessment model used in this report; (3) development of plans to meet these needs; (4) monitoring of petroleum production, other economic activities, and their effects on shore resources and the environment, with a budget provided jointly by industry and government.

Some definitional problems will have to be decided quickly by the commission in consultation with industry, all levels of government, the native and resident organizations, and public bodies such as the Canadian Arctic Resources Committee (CARC) and the universities. These problems include a workable definition of the shore zone; which land use, resource, environmental and other effects are to be monitored and why; and what arrangements are necessary to avoid duplication with any monitoring programmes in existing government agencies.

The shore zone commission should be composed of members with varying backgrounds, including several with both management and research experience. The membership should be representative of all major interest groups and should have a full-time chairman as well as an adequate full-time support staff and budget.

As part of its research and monitoring duties, the commission should begin immediately to receive and comment upon all land and marine proposals likely to have a significant effect on resources and the environment in the shore zone, prior to their approval by any federal, territorial, or local government agency. A procedure will have to be worked out for determining what is a significant activity and, in particular, for commenting upon and monitoring apparently insignificant projects which could have important medium- to long-term cumulative effects. It will also be necessary to watch carefully for proposals that are likely to undergo major modifications, as these could have large unforeseen impacts. In commenting on proposals on the basis of its research and monitoring functions, the commission would often provide recommendations on desirable courses of action for oil and other developments, as well as protection of existing land uses and the resources and the environment upon which they are dependent.

The need for a body such as the Beaufort Sea-Mackenzie Delta shore zone commission is urgent. Crucial decisions are already in the balance — for example, the desirability and feasibility of concentrating port, oil terminal, refinery, servicing, and other activities in a particular area or areas. The shore zone commission can also assist with some of the pressing planning and management functions — such as those currently performed in the Shetland Islands by organizations such as the harbour authority and the council — at least until more suitable permanent arrangements can be planned and implemented.

5. Create an Arctic Environmental Forum

Creation of an Arctic Environmental Forum would provide a focus for exchange of information on petroleum and other new economic developments, their likely effects on pre-existing activities, and the resources and environment upon which they are dependent. The forum should be a mixing place for all interested parties: industry, all levels of government, native people, other local residents, southerners, university or other experts, and knowledgeable concerned individuals. The forum should meet at least twice a year and should hold a conference on the arctic environment at least bi-annually, preferably with the Department of the Environment, industry, and major citizens groups such as CARC and the Association of Canadian Universities for Northern Studies (ACUNS) as co-sponsors, along with local land use and conservation organizations. The forum should be financed jointly by the federal government (Department of the Environment) and industry. This forum could be affiliated with the existing Canadian Environmental Advisory Committee (CEAC). If this affiliation were to be pursued, the forum's operations should be to a considerable degree free of the wide-ranging responsibilities of and demands placed upon CEAC.

6. Improve Canada's shore zone co-ordination system

The shore zone co-ordination system recently developed by the federal, provincial, and territorial governments as a result of concerns documented at the 1978 Canadian Shore Zone Conference in Victoria should be expanded considerably as soon as possible. The process should begin in the federal office with three basic studies conducted in association with the regional co-ordinators at the provincial and territorial levels:

1. Assessing shore zone research and information systems and their efficiency and effectiveness, particularly for the Beaufort, the Maritimes, and other areas under intensive oil development or other pressures; comparing these systems to other potentially instructive models in other parts of the world, for example the North Sea countries and the United States; and making recommendations to interested parties both within and outside of government.

2. Assessing the efficiency and effectiveness of existing methods of co-ordinating the responsibilities and activities of various levels of government in the Canadian shore zone, particularly the methods used in areas under heavy development pressure; comparing these with potentially instructive models elsewhere in the world; and making recommendations to interested parties within and outside of government.
3. Assessing the efficiency and effectiveness of offshore or marine management systems in various parts of Canada, particularly areas under intensive oil development pressures, and comparing them with potentially instructive models in other parts of the world. Offshore or marine management systems include those between about three miles offshore and the limit of the 200-mile economic zone, where Canada has recently acquired responsibilities for managing an array of pre-existing and emerging national and international uses — including oil development, mining, fishing, transportation, navigation, and defence. This is the sea use planning and management problem currently of increasing concern in the United Kingdom and the United States. It involves multiple-use management not only of offshore activities, their conflicts, and effects, but also of their impacts upon nearshore uses, resources, and the environment.
4. In developing to provide needed information on Canadian shore zone management generally, the federal shore zone co-ordinator's office should be expanded. It should include a library (data processing), and a clearinghouse function for the data being published in many parts of the world on the increasingly complex problems of what is essentially a new coastal and marine frontier. It would also be advantageous to appoint a Canadian Shore Zone Advisory Committee to meet at least twice a year to identify issues, suggest significant research, and recommend suitable approaches based on the experience and knowledge of its members. The committee should consist of both government and non-government members in about equal proportions.

Some of the foregoing recommendations are rather novel, while others may be seen as "old chestnuts" — for example, the land claims and local and regional government issues. What is new and vital about these in the context of this report is that their significance can now be seen in relation to a successful example — the Shetlands. Basically the recommendations provide for an indigenous thrust in the Beaufort — a balance against the heavy pressure from outside. The meeting of these two forces is much more likely to produce effective development and resource and environmental management in the northern coast and offshore waters than is the present incomplete and fragmented structure.

In-filling and new service facilities at Lerwick harbour, Shetland Islands, September 1980, J.G. Nelson

Appendix 1: Questionnaire or interview guide on applicability of the Scottish shore experience to Canadian Beaufort Sea development

We have been asked to undertake a study and make recommendations for improved shore zone management in the Beaufort Sea area on the basis of experience in Scotland, including the Shetland Islands. The project is being co-sponsored by the Department of the Environment and the Department of Indian Affairs and Northern Development.

As an aid in our study, a management assessment model or analytical framework has been developed (see Table 1 in this report). This framework was developed from pipeline and other studies in the North and an evaluation of the environmental regulatory system for a multi-billion dollar greenfield industrial complex in Ontario. It has been modified slightly for the present study.

The management framework is intended to describe the major management steps or information categories that are considered necessary to make sound decisions on large industrial and technical developments and their impacts on the social and biophysical environment. It is not intended to describe the sequence in which these steps will occur as these may vary with the situation. Furthermore, much looping or feedback may occur among the categories as management proceeds.

The meaning of the framework is set forth more fully in the comments and questions in the attached questionnaire. We would like to request that you complete the questionnaire and also send us any other comments you may have by 20 August 1980. Some of the questions could take considerable time to complete. Your own resources and commitments may only permit short, direct responses, perhaps on the questionnaire itself. In completing the questionnaire you may only wish to address certain items because they are relevant and inside your area of expertise. At any point where it seems appropriate, please list any reports or publications which might help us and provide other contacts.

It is important to note that the questions are often stated in terms of experience with a particular agency, corporation, or group. If you are not replying on this basis, but rather through different or more varied experience as an individual, please so indicate and then treat the questionnaire *as a guide* in structuring your answers. If you wish the response to be confidential and used without specific attribution would you please so indicate.

We are very grateful for any assistance you can give us with this project. On the basis of contributions such as your own, we hope to prepare a report which will aid in improving management of the Beaufort Sea.

Agency

1. What is your mandate(s) in the Beaufort Sea? Is the mandate(s) set forth in legislation, in policy, or in other ways? Could copies of relevant acts, policies, or other documents on mandate be provided to us?

2. Is your agency or group a *lead* agency in regard to any environmental or social (includes economic, political, and cultural) aspects of Beaufort Sea development? Is your agency or group a *participating* agency in regard to any aspects of Beaufort Sea development? Briefly, describe the agencies with which you are involved in both the above contexts; a listing will suffice. Detailed discussion of committee or other forms of co-ordination and their advantages and disadvantages in terms of experience or theory can be discussed here or later under question 15.

Planning

Strategic planning

3. Has your agency (corporation or group) carried out strategic planning in regard to environmental and social aspects of Beaufort Sea development? In this context has your agency set specific goals and objectives for attainment, and criteria to judge success? Has a basic management philosophy — multiple-purpose use, for example — been set forth and defined for interested groups? Have management steps and schedules been established — for example, inventories, land or water classification systems, and other components as noted in Table 1? Can we be provided with any documents describing your strategic planning? Are you aware of strategic planning by other agencies as it affects your activities? Were you involved in their planning? Do you have comments on this planning?

Inventories (Biophysical)

4. Has your agency conducted biophysical (biological, geological, etc.) inventories for oil and gas exploration, pipelines, or other developments in the Beaufort Sea area? Could we have copies of such inventories? Are you aware of inventories done by others? Do you use these? Can you comment on the advantages and disadvantages of such inventories?

Inventories (Social)

5. Has your agency conducted social (economic, political, or other studies which can be defined as broadly social) inventories in the Beaufort Sea area? Could we have copies of such inventories? Are you aware of inventories done by others? Do you use these? Can you comment on the advantages and disadvantages of such inventories?

Land classification

6. Have you developed or used any land classification systems based on inventories like the foregoing or other research, and have these systems been used in reducing land use conflicts or land use impacts on the environment or people by guiding the spatial location of exploration or other activities? Can we be provided with copies of these systems, information on their use, advantages and disadvantages, and other relevant details — for example, evaluations of the systems? Are you aware of such systems being used by other agencies (corporations or groups) and do you have comments on these?

Especially significant areas

7. In light of the foregoing, have any land or water classification categories been developed which identify special land use or environment zoning? These zones or areas have been given various names in other jurisdictions — for example, environmentally sensitive areas or special treatment areas. Such zones include calving areas, spawning grounds, marshes and wetlands, archaeological and historic sites, or other areas of special biophysical or cultural importance. If such zones have been identified may we receive copies of reports or other information on the classification systems involved, any policies or management guidelines associated with use of such zones, and any review reports, or information on their advantages and disadvantages or other relevant details? Are you aware of other uses of environmentally sensitive areas or special treatment classification by other agencies (corporations or groups) and do you have comments on these?

Evaluation

8. Has your agency (corporation or group) conducted any evaluation of policies, programmes, technology, or practices as part of its management activities in the Beaufort Sea? We are thinking here of formal evaluations which have been given different names according to their focuses. We include

policy evaluations, technology assessments, environmental impact assessments, social impact assessments, and benefit/cost evaluations. If such evaluations have been conducted may we have copies? If they contain confidential information, summaries or other indications of the nature of such studies would be appreciated. Would you also indicate how frequently you use such evaluations and under what circumstances — that is, their timing, and who is involved or consulted? Do you have policies or guidelines in regard to such studies? If so, may we have copies? Can you comment on how these evaluations are used in and effect management? Are you aware of the use of such evaluations by other agencies (corporations or groups) and would you comment on such use?

Implementation

Approvals

9. Although no formal classification of the different means of securing approval for exploration or other economic or land use activities has been developed, two broad types seem to be used and recognized. The first attempts a relatively comprehensive examination of activities, technology, and environmental and social impacts prior to giving permission to proceed. It is probably best exemplified in the U.S. — for example, in the securing of permission to drill on the Alaskan North Slope. The second is a negotiation process and generally involves close consultation between industry or other proponent and government approval bodies throughout the planning process. This latter approach appears to have been more commonly used in Canada, although it has been attracting more attention lately in the U.S. where it is becoming known as "negotiated development."

Would you comment on the major advantages and disadvantages of these two systems and offer suggestions as to which modifications or other changes might improve the approval process from the standpoint of any agency (corporation or group) like your own? In thinking about your answer you might wish to consider the pre-hearing, hearing, formal case record and other procedures identified as desirable in various studies of regulation supported by the Conservation Foundation in the U.S. You might also wish to consider criteria such as efficiency, effectiveness, equity, costs, and benefits. Are you aware of other processes used elsewhere and can you comment on these in terms of their suitability for the Beaufort Sea or comparable areas?

Construction and operation

10. Once construction begins, what surveillance and inspection mechanisms does your agency (corporation or group) use in order to ensure compliance

with conditions, specifications, or approval understandings for management of social and environmental impacts? Do you have codes, guides, or other documents for use by inspectors? Within what unit in your agency (corporation or group) are the surveillance and inspection personnel housed, and to whom do they report? What professional or other background do you seek in inspectors and how are they trained by your agency (corporation or group)? What do you see as the advantages and disadvantages of your surveillance and inspection system and how might it be improved? Are you aware of surveillance and inspection systems for other agencies, corporations, and groups involved in the Beaufort Sea and would you comment on their advantages and disadvantages and any other aspects you consider relevant?

Monitoring

11. Monitoring is difficult to distinguish from surveillance and inspection and, as will be seen later, from research as well. Monitoring can be defined as a more continuous and somewhat higher level of activity than surveillance and inspection. Monitoring in effect is a way of evaluating surveillance and inspection as well as the various conditions, specifications, and understandings reached at approval time. Monitoring can therefore involve continuous measurements of air and water quality, stream sediment load, vegetation regrowth, housing prices, crime rates, unemployment rates, and other environmental and social factors.

Does your agency (corporation or group) conduct such monitoring? If so, please describe briefly what is monitored and why, and how the results are taken into account in management, giving examples if possible. Is monitoring an administratively recognized activity in your organization? If so, where is the activity placed in your system and to whom do the monitors report? May we please have copies of any monitoring reports and any documents indicating action taken on them? Are you aware of monitoring by any other agency (corporation or group) involved in the Beaufort Sea and do you have comments on the advantages, disadvantages, or other details pertaining to such monitoring? Do you have any comments on the idea of an independent monitoring unit jointly financed by the developing companies and government and on the idea of an information centre supported in a similar way?

Enforcement

12. What enforcement systems is your agency (corporation or group) subjected to or practising in regard to meeting approval conditions, specifications, and understandings? Could you comment on the advantages

and disadvantages of these systems? Are you aware of the enforcement systems applied to other agencies, corporations, or groups involved and would you comment on these in light of the previous questions or other matters you consider relevant?

Modification

13. Modification may be necessary to conditions, specifications, or understandings when unexpected or changed circumstances are encountered during construction or operations. Does your agency (corporation or group) have policies, guidelines, codes or other procedural systems for managing similar modifications which may be of a relatively minor or major nature, such as sensitive completion of stream crossings or relocation of a road or other facility? One important form of modification is the closing down, or re-fitting of a plant or other operation as mineral bodies deplete or other factors change. Rehabilitation procedures may also be necessary or desirable towards the terminal stages of a project or programme.

To what extent and in what ways are such modifications taken into account in strategic and other planning for projects and programmes in the Beaufort Sea or similar areas? Is the current planning and implementation system too static in regard to such modifications? Are you aware of the ways in which such modifications are managed by other agencies, corporations, and groups, and what are the advantages? Do you have any other comments on this item?

General guides or characteristics

Research

14. As a result of the technical, social (includes economic, political and cultural) and environmental (biophysical) complexity of developments such as those underway in the Beaufort Sea and similar areas, research is increasingly seen as a continuing and necessary part of sound management. Indeed, many of the management activities set forth in Table 1 and referred to earlier in this questionnaire depend on the extent and quality of research conducted by the agencies, corporations, and groups involved. Would you therefore provide a general description of your research operations and the ways in which your research is co-ordinated with other involved agencies (corporations and groups)? We are particularly interested in the research conducted on the social

and environmental impacts of oil and gas exploration and development, including associated or consequential activities and effects. Does your agency (corporation or group) conduct social as well as environmental research? Can you provide us with examples of such research and cost and other data relating to this research? Would you comment on the advantages and disadvantages of research as it is undertaken by your agency (corporation or group) and can you offer suggestions for improvement? Are you aware of research programmes of other agencies, corporations, or groups involved in the Beaufort Sea and similar areas and would you comment on the advantages and disadvantages of these programmes?

Co-ordination

15. The extent and nature of co-ordination has been a major concern in complex situations where many government, corporate, and private groups and interests are involved. Concern has been expressed about overlap, redundancy, inconsistency, lack of clarity, and other factors relating to efficiency, effectiveness, equity, costs, and benefits in the Beaufort Sea and similar areas.

What co-ordination activities has your agency (corporation or group) been involved in and how have these activities been organized? If committees or other mechanisms are involved, what are their advantages and disadvantages? What should be done to improve the situation? In addressing these questions, please consider the various types of co-ordination noted in Table 1 — that is, among government agencies, among industry, among public groups (such as non-governmental organizations or environmental or conservation groups), and also among these groups (inter-group). In this regard, it has been suggested that impending development in the Beaufort Sea area will necessitate some overall co-ordination group or a higher-level integrating decision-making body or system. Regional planning councils, special agencies or authorities, and a special shore or coastal zone management system are among the mechanisms being considered. What are the advantages and disadvantages of these mechanisms? What is your own preference under the cirumstances and why?

Economic incentives

16. Regulations have been seen as unduly numerous and complex and also as an ineffective and costly way of managing Beaufort Sea and similar developments. Do you believe that economic incentives or approaches would be more efficient and effective as well as equitable? What are your views on the way economic incentives have been provided in the Beaufort Sea situation,

with regard not only to your agency (corporation or group), but also to others with responsibilities or operations in the area? Do you have any suggestions for improving the use of economic incentives and other steering mechanisms or guides in the Beaufort Sea area?

Information access

17. The flow of information on all aspects of Beaufort Sea development has been a source of much concern, especially among public interest groups. What are your views on this matter? What kind of information should not be released and why? How can the information flow system be improved with respect to the nature and quantity of material, the timing of release, the opportunity for comment, and so forth? Can various kinds of joint screening bodies or other mechanisms be established which would improve the situation? What are their advantages and disadvantages? To what extent would freedom of information or comparable legislation or legislative amendments be helpful? To what extent are the perceived difficulties organizational or financial in nature — that is, a result of inadequate information dissemination centres and systems, or inadequate co-ordination and funding by government or other bodies?

Management process monitoring

18. This is conceived of as high-level monitoring of the entire management system and process by the responsible agencies in association with independent centres or groups. Such a mechanism was suggested for the Alaska Highway pipeline and has been used at least in part in a recent review of the Great Lakes water quality agreement between Canada and the United States. The intent is to follow, evaluate, and report on the entire programme, including: goals and objectives; criteria for judging success; technical, institutional, or other means of reaching these; planning systems including inventories, and classification systems; adequacy of inspection and co-ordination; and other categories outlined in the management framework noted in Table 1. Such a management monitoring process could avoid perpetuation of inefficient, ineffective, inequitable, and costly programmes which might continue without careful independent review and constructive advice. What are your views on such a procedure and on possible ways of organizing it, if it proved desirable? What are its advantages and disadvantages from your standpoint? What alternatives are open in this regard?

Comments

19. Do you believe the analytical framework for management noted in Table 1 and expanded upon in the foregoing questions is a useful one? Why? Would you add any factors to the model or can it be clarified in any way? What are its advantages and disadvantages in the context of the Beaufort Sea and similar areas? Please add here any comments on management issues, problems, and possible approaches that you wish to make which you do not believe are adequately covered in the foregoing questions or in Table 1. This is also a good place for afterthoughts on any of the questions or items listed in Table 1.

Point of clarification

Subsequent to the preparation of this questionnaire, it was decided to concentrate on the land use, resource, and environmental aspects of the Scottish and — to a lesser extent — the Alaskan experiences. As a result, although there are references to social aspects and effects in the questionnaire, these matters are not considered in detail in the report.

Appendix 2: Response to questionnaire from the Canadian Arctic Resources Committee

Agency

1. The Canadian Arctic Resources Committee (CARC) has no formal or official mandate in the Beaufort Sea region, but it does have a self-assigned mandate that requires the committee to examine the engineering, economic, environmental, and social implications of development in the Beaufort-Delta region. This broad mandate was first assumed by Doug Pimlott as a field worker in the Delta in the early 1970s and has continued through CARC's publishing programme and specific research projects since then. In this respect CARC works very closely with both the Committee for Original Peoples' Entitlement (COPE) and the Dene Nation. Publicly, CARC is seen to be a principal critic of Dome Petroleum and of government performance in the area.

2. CARC is a lead agency only in the sense that it is probably the lead non-government organization with respect to arctic matters (both environmental and other). In fact it was CARC that really established itself as the outside authority on offshore drilling, beginning with Doug Pimlott and the publication of the book he co-authored — *Oil Under the Ice*.

In doing its Beaufort work, CARC interacts principally with government agencies in Ottawa and Yellowknife, Dome Petroleum in Calgary, and COPE in Inuvik. However, in the past there has been substantial involvement with the other principal developers in the region (Esso Resources Canada, Shell Canada, and Gulf Canada Resources).

Planning

Strategic planning

3. While CARC agrees that multiple use for large parts of the Beaufort-Delta region may be quite proper, we do not believe that this can be applied to all the land mass or water areas. In this respect CARC has been very concerned with what it calls "balanced" development. This means that CARC has urged government, in the face of the often overwhelming pressure of industry, to

carefully evaluate all the resources of the region to ensure that those that require protection will not be prejudiced by the encroachment of non-compatible uses (principally in the form of oil and gas exploration and development). In this respect CARC has suggested that it would be wise to designate or survey the environmentally significant areas, and the culturally significant areas, to designate lands that will be required under a land claims settlement (whether they be developed or not), and to otherwise set up the management infrastructure that will permit rational decision-making both in the local and national interests.

We are advised that the Department of the Environment has recently completed a survey of environmentally significant lands that come under its mandate (Canadian Wildlife Service, Parks Canada, etc.) and we would assume that this includes the Beaufort-Delta region. Nothing has been made public to date.

In addition, we have three other documents that indicate that comprehensive strategic planning in the North is gaining momentum. The first document is the Department of Indian Affairs and Northern Development's *Direction for the 80s.* Although this document has not yet received official approval by the Minister, it is the document that now guides departmental programmes. The second document is a draft Cabinet memorandum on northern land use planning. That draft is now being revised and it is expected that it will emerge sometime in 1981 to guide regional land use planning in the entire North. The third document is a Cabinet memorandum on the approach to comprehensive native claims.

Inventories (Biophysical)

4. CARC does not conduct any independent biophysical, geological, or other inventories. Rather it relies on assembling this information from other sources. We use information published by industry, both through industry journals or other publications, and from documents that are otherwise made available to CARC. We also assemble data from all government departments and try and synthesize this to produce a sensible picture of what is actually going on on the ground. The difficulty is that the information from any one source is highly suspect and in many cases is deficient in that it seeks to justify a certain end rather than to provide an objective overview of the situation.

5. CARC has conducted a number of studies of the Beaufort Sea area, and most of these are freely available in our publications, such as *Oil Under the Ice,* various articles in *Northern Perspectives,* etc.

We are always on the look-out for inventories that are done by other agencies and we use these whenever possible. The difficulty is that many of these inventories or studies are not freely available and are not part of the literature that researchers or public interest groups can normally get their hands on.

Land classification

6. The principal land classification system that we are aware of for the Beaufort-Delta region is the one that arises from the COPE Agreement in Principle. We have suggested on many occasions that this be expanded upon and made more detailed, but to our knowledge there is very little else that is available at this time. Of course there are particular land classifications on a site-by-site basis (Kendall Island Bird Sanctuary) but these are not knit together in an overall framework that makes any sense from a local, regional, or national perspective.

Especially significant areas

7. As noted in the previous answer there are a number of specific land and water classifications that have been developed that identify special land use or environmental zoning, but the fact remains that these have either not reached maturity or have never been knit together to make a sensible whole. For example, there are a number of areas suggested under the COPE Agreement in Principle; Judge Berger made recommendations for whale sanctuaries and a number of other environmentally sensitive areas; and the Department of the Environment has various ideas for parks, national wildlife areas, and so on. The problem of lack of definition of these areas on a map is further compounded by the lack of management guidelines that govern the use of areas that are already acknowledged to be important. CARC would certainly advocate the type of overview study that was done by the University of Waterloo for the Yukon.

Evaluation

8. CARC has concentrated a great deal of its time and energy and its publishing programme on the evaluation of policies, programmes, technology, and other practices in the Beaufort Sea. This began principally with the book *Oil Under the Ice* and has been followed up since in a number of issues of *Northern Perspectives* and other studies that CARC has commissioned for its own use. All such studies, if they are not already published, are freely available from CARC's files.

The use of such evaluations from a management perspective is something that is subject to some speculation. However, one can safely say that because of the CARC studies, there have been a number of shifts in government and company performance in the Beaufort-Delta region. Certainly the most recent example would be the study that CARC did on dredging in McKinley Bay (published in *Northern Perspectives*, Vol. 8, No. 2). This document of how the government decided with respect to major activities in the Beaufort-Delta region, proved to be an embarrassment to both the Department of the Environment and the Department of Indian Affairs and Northern Development. We have been advised by officials in both departments that this exposure was one of the reasons that the production systems in the Beaufort-Delta region have been referred to the Environmental Assessment Review Process.

So the publication of critical studies does have an effect. The effect is often difficult to pinpoint right away, but in the long run one can safely say that it does influence government performance and policy and that it certainly affects the performance of industry. Gordon Harrison, the principal officer of Dome Petroleum responsible for the Beaufort Sea, has said to CARC recently that the early work done by CARC did in fact benefit the company in that it made them proceed much more cautiously.

Implementation

Approvals

9. The various types of approval processes have advantages and disadvantages. The difficulty that I can see from the information that is provided is that one cannot divorce the approval system from the overall decision-making framework — including the *political* framework within which these approval systems are operating.

The classic example is the difference between Alaskan North Slope approvals and the Beaufort-Delta approvals. Certainly the elaborate constitutional and legal framework that supports the interest of individuals and wildlife conservation on the Alaskan side of the Beaufort Sea does not exist on the Canadian side. For example, we have no endangered species act to protect the bowhead whale, and yet such an act in the United States is being used in the state and federal supreme courts to modify and/or halt offshore drilling in the Alaskan Beaufort.

Clearly there is an advantage to the U.S. system of taking a general overview (by government EIS) of the total context of what is being planned in the short- and long-term and then, after that assessment is made, to proceed in a step-wise fashion according to the types of analysis or impact assessment that have been agreed upon. This of course implies that the overview is conducted by someone other than the industrial proponent. The government accepts this responsibility in the United States but it does not accept it in Canada. To try and change this would require a considerable upset of the current thinking of government. I do not think that this is realistic in the short-term. This is not to say that we should not work towards such change and perhaps the federal government's Lancaster Sound green paper exercise is the beginning of such a change.

Tied into this directly is the whole role of, and responsibility for, research. Of course if the industry is going to be the proponent and first evaluator, then the industry will retain the research initiative and government will simply be in a position to react. The difficulty is that the capability of government to react decreases the longer this type of situation is allowed to perpetuate itself. Certainly there is ample documentation now to show that the government's research capability in the Arctic is critically low — so low in fact that many people will argue (and I will certainly support them) that the government no longer has the capability to assess the initiatives of industry in the Arctic.

Construction and operation

10. CARC, of course, does not get involved directly in construction and operation of projects and hence does not get officially involved in surveillance and inspection. However, CARC does monitor these projects from a distance. When we are able to obtain information, it is analyzed, and we seek justification and/or rectification of the problem. For example, we are now doing this on the Alaska Highway gas pipeline. Information is sent to us by a number of sources (some known and others unknown) and we then take this information and through our own publishing programme, and bureaucratic or political connections (with the use of the media in certain circumstances), we seek to have the situations addressed. The real dangers that we face are that we get off on tangents or that we do not have all the information. However, once aware of the dangers it is easier to avoid them, and we have managed to achieve remarkable success on the issues we have tackled.

Certainly with respect to the Beaufort Sea one can point to a number of studies (many of them done by CARC) that indicate that the surveillance and inspection mechanisms are perilously inadequate. For example, see the recent edition of *Northern Perspectives* (Vol. 8, No. 6) in which a letter from the chief conservation engineer in Yellowknife to the minister outlines the fact that the number of oil and gas inspectors is far below that necessary to assure adequate control of the industry's operations in the Beaufort. The same point was raised at a court case in Yellowknife in 1980 and has arisen time after time since drilling began in the Beaufort Sea.

Monitoring

11. Monitoring in the Canadian system is almost non-existent. The very little bit that is done and reported upon simply does not feed itself back into more effective surveillance and construction practices. Here I am thinking principally about some of the monitoring that has been done by the Department of the Environment on offshore oil and gas development. When the results and recommendations of this work were published the reaction from Energy, Mines and Resources and DIAND was extremely strong and it resulted in a bureaucratic fight that was entirely counter-productive.

Certainly one of the most effective monitoring mechanisms that we have seen — at least in embryo form — is that of the Alaska Highway Pipeline Panel, and the Environment Protection Board. If these types of institutions had carried on past the assessment stage they would have benefited not only the public in its perception of a project but also those closely involved with the project itself.

The idea of jointly financed monitoring agencies is one that has merit for the book-keeper mentality, but it is flawed in that it begins to mix up the respective roles of industry and government.

Enforcement

12. The big-stick approach to enforcement is not commonly used in Canada, if at all. Canadians seem to pride themselves on the fact that they reach negotiated settlements. Certainly this has been my personal experience sitting as a member of the Northwest Territories Water Board. It was only after many, many years of the most blatant abuse of effluent standards that the board finally recommended that a mine be considered for court action. This initiative was stalled by both the Department of Indian Affairs and Northern Development and initially by the Department of the Environment. It was only when the situation became widely known to the public and clearly embarrassing to the departments that the government decided to take this issue to court. The same situation can be seen in the Beaufort Sea where there have been consistent and significant violations of standards and procedures.

I say this recognizing that not all enforcement should be something that we see in the courts. Clearly the most effective enforcement is something that involves working with the developer on an ongoing basis to try and reach mutually acceptable solutions to a problem. However the virtual absence of any vigorous move by government can only be interpreted by the public (and indeed by the proponent himself) as a licence to proceed with much more relaxed standards than one would otherwise assume are being implemented.

Modification

13. The modification of a situation or a proposal is one that, in CARC's view, must be incorporated very early on in the planning process. If this is not the case, particularly with respect to multi-billion dollar projects, it is virtually impossible to seek an effective solution to the problems that ensue. For example, in the Beaufort Sea the planning that Dome is now working with involves the transportation of the hydrocarbons by tanker through the Northwest Passage. If this is not assessed quickly and if decisions are not reached soon, then the company's planning will take them to the point of no return. Inevitably it will result in terrific political pressure to make sure that the decision-making framework on the government side concurs with the company's plan. It is therefore imperative that government decision-makers be in a position to reach decisions early in the planning process — even in the pre-planning process — so that industry's resources can be directed properly and that its project can be modified in its early stages. If this capability existed, then in my view, the vast majority of conflict situations that do arise in the North would be avoided. Of course the problem is that the government decision-making framework is so lethargic and/or non-existent that industry is left with an entirely free hand until the very last minute — at which time

government starts to have second thoughts and/or begins to impose additional conditions that were never part of the "deal" in the first place. The example of Norlands Petroleums in Lancaster Sound is a classic horror story for industry.

General guides or characteristics

Research

14. CARC has no specific research programme with respect to the Beaufort Sea that ties together all the various activities that we engage in. This is not particularly alarming, because the committee has directed that the Beaufort Sea is a continuing priority of the committee. Hence, it forms the background against which much of our other studies are conducted.

For example, if we are doing non-Beaufort Sea research and a situation arises that requires the diversion of more resources to the Beaufort Sea, then that is automatically done. A rough estimate of the financial costs of Beaufort Sea research would be in the order of $15 to $20 thousand a year by CARC. It may be more or less in any given year depending on the level of activities of government and industry in the region. What this figures does not show, however, is how other research that CARC undertakes also relates to the Beaufort Sea. For example, the study that CARC did on the Environmental Assessment and Review Process looked extensively at particular operations in the Beaufort Sea and used them in case studies for the evaluation (see Bill Rees's analysis of the McKinley Bay dredging).

CARC has recently begun a research programme on the government's arctic research. This was initiated because our work began to indicate that the government's capability to conduct research (and hence evaluations) was being put in serious jeopardy. For example, we found that the Arctic Biological Station at Ste Anne de Bellevue had a staff cut last year of twenty per cent and will have a further staff cut of twenty per cent. In addition, the budget for research at that station has been systematically cut over the past several years until now it is virtually non-existent. The same kind of thing can be seen elsewhere. Combined with the loss of financial resources is the loss of professional resources. If one were to compile a list of scientists with particular expertise in the Arctic who have left government service in the last five years, I would suspect it would make a rather impressive document. This kind of pattern is all too common and the result tends to amplify the problems that are being created between industry and government in the capability to do and assess research.

Co-ordination

15. Perhaps one of the least-understood and least-publicized activities of CARC is co-ordination and synthesis of information. Drawing together information from a variety of government departments and other sources,

synthesizing it, and pushing it back out in the form of publications, and letters to various individuals in industry, government, and other places, helps the information flow — which in our view is one of the key elements of co-ordination. The situation becomes absurd in certain circumstances when we find that officials in one government department or another will write to us and provide us with confidential information, so that we can then press the minister or some bureaucrat in that or another department, or somebody in industry, to dislodge something that is holding up progress. This is a risky business in that CARC can often be used by government agencies or by industry, but it is one that is becoming increasingly common. Quite apart from the details of the role here, it does indicate that the co-ordination in government is bad.

There is no question in my mind that some sort of regional planning council or special agency is necessary for the Beaufort-Delta region. It should be locally based and should seek to have real authority vested in the local people with only an overriding authority where issues of national interest are concerned. Of course the inherent difficulty in setting up something like this is that it requires the release of power from bureaucrats at the federal level to people at the local level. This is not easily achieved and, from my own experience with the bureaucracy, is impossible to achieve without a bureaucratic revolution. In a sense, the COPE claim may turn out to be the catalyst that causes this revolution, but it is certainly not going to be easily or quickly won.

Details on how a regional approach might be taken have been tossed around by CARC and COPE and others for over a year and a half. Draft papers have been prepared and a conference is being organized for sometime early in 1981 to address this particular problem.

Economic incentives

16. There is nothing wrong with regulations *per se,* but there is something wrong when regulations become an *ad hoc* way of providing a planning framework. I think one could effectively argue that a sensible regulatory framework could only be put in place when one understands the context - the planning context — within which such a framework would be situated. The reality of the situation in the North is that various acts and regulations are promulgated over a period of time in response to different types of political and other pressures. The result is a plethora of individually sensible, but collectively irrational, committees, advisory groups, boards, and other kinds of regulatory mechanisms. This is now being recognized within Indian Affairs and Northern Development, and there seems to be work underway to make some sense of the nonsense that has been created. Certainly the draft cabinet paper on regional planning indicates this, and rumours of a complete revision of all environmental standards for the North are becoming increasingly common.

Specific mention is made of economic incentives and certainly this is one area that has to be considered when trying to understand the Beaufort Sea. In the past, with the superdepletion allowance, the economic incentive for industry has been absolutely overwhelming, and when properly understood, embarrassing for both government and industry. No doubt this is partly why the government has taken such a drastic measure in the new energy budget. However, even the budget provides incentives that — in the words of Bob Blair — absolutely compel the industry to go North.

And the economic incentives cannot be viewed by themselves. What they do is distort the total planning framework so that the industrial activities are much better financed and much further ahead than any other segment of the system that exists in the region. For example, if Dome Petroleum had not been given the superdepletion allowance, the offshore oil and gas development would have taken place at a completely different pace, probably one much more compatible with the territorial levels. The introduction of a terrific stimulus to one segment of the economy in one region requires the government to provide compensation financing for other parts of the economy so planning and other requirements can keep pace.

Information access

17. There is no question that information is the key to proper planning, assessment, understanding, and administration. And there can be no question that information access for activities in the Beaufort Sea has been grossly inadequate. The ultimate solution is unquestionably a freedom of information act or comparable legislation. However, in the meantime, the shifting of decision-making power away from a highly centralized, and isolated bureaucratic environment in Ottawa, and perhaps even in Yellowknife, would go a long way towards helping. There is nothing better than real political checks and balances that provide real political power at the territorial and regional levels.

Management process monitoring

18. I think that the Alaska Highway Pipeline Panel is perhaps the best model for effective monitoring that we have had so far in Canada. The difficulty is that the success of the panel has had as much to do with the personalities involved as with its structure. Of course this is true for everything, but it is even more true when one is dealing with the terrific imbalance in resources between industry, government, and monitoring agencies.

Comments

19. I think that the analytical framework that you have presented is a useful one for analysis and will provide very good insights into the strengths and weaknesses of the management systems we have evolved over time. However, I would add one word of caution. The analytical framework you have set up

bears great resemblance to some sort of computer programming approach to decision-making. This, while it may be ideal on paper, does not in fact exist in any institution that I know of. It certainly does not exist in government in any way, shape, or form. Therefore the danger in adopting this is that you have created a standard against which to evaluate something that may have no bearing on what is actually happening. Decision-making in complex organizations and on complex issues is, as much as anything else, the art of creatively muddling through the problem. One need not go much further than decision-making in their own personal life to see this. If we try and imprint a rigid, methodological, analytical, cause/effect type framework on this, the system cannot respond and certainly cannot be understood as it is. What this boils down to is that we have to accept that a large part of decision-making is not entirely rational. It is based as much on feelings, emotions, and external pressures as it is on management theory, economics, and facts. Hence the kind of conclusions one would reach using the model put forward will be exceptionally useful and provide clear insights where none have existed before, but it will not provide the total picture. In effect it will point us in the direction we should be going; but at some point we have to leave all models and simply propel ourselves in the direction that they point us so that we can "muddle through" in a more creative fashion.

Appendix 3: Interviewees and respondents: Scotland, Canada, and the United States

A. Major consultations regarding Scotland, 1980

Ross Atkins, Department of Geography, University of Aberdeen.

Jennifer Baker, Research Director, Orielton Field Centre, British Field Studies Council, Pembroke, Dyfed, Wales.

Ron Bisset, Department of Geography, University of Aberdeen.

Keith Chapman, Department of Geography, University of Aberdeen.

Brian Clarke, Department of Geography, University of Aberdeen.

Frank Dean, Chief Environmental Planning Officer, British Gas Corporation, London, England.

Professor G. M. Dunnet, Department of Zoology, University of Aberdeen.

Captain Albert Flett, Director, Sullom Voe Harbour Authority, Shetland Islands.

Ian Townsend Gault, Research Associate, Canadian Institute of Resources Law, University of Calgary.

Alaistair Grubb, Oil Pollution Control Officer, Sullom Voe Harbour Authority, Shetland Islands.

George Hunter, Shetlands Fisherman's Association, Lerwick, Shetland Islands.

Tony MacKay, Institute for Study of Sparsely Populated Areas, University of Aberdeen.

Gordon Mann, Director of Planning, Shetland Islands Council, Lerwick, Shetland Islands.

David Monk, Environmental Officer, Sullom Voe Terminal, Shetland Islands.

Glenys Parry, Nature Conservancy Council, London, England.

Jeremy Rown-Robinson, Land Economist, University of Aberdeen.

John Smith, Department of Geography, University of Aberdeen.

Peter Suttie, Planning Director, Banff-Buchan District Council, Banff, Scotland.

Peter Wyllie, BP Representative, Lerwick, Shetland Islands.

B. List of questionnaire respondents: Canada

Lloyd Brooks, Victoria.

R. Clarke, Fish Habitat Section, Department of Fisheries and Oceans, Ottawa.

A. T. Davidson, Parks Canada, Department of the Environment, Ottawa.

Don Gamble, Canadian Arctic Resources Committee, Ottawa.

Michael P.G. Hawkes, Resource Planning Branch, Department of Renewable Resources, Yukon Territory, Whitehorse.

Manfred Hoefs, Wildlife Branch, Government of Yukon, Whitehorse.

Dave Mossop, Wildlife Branch, Government of Yukon, Whitehorse.

Michael Parkes, National Co-ordinator, Shore Zone Integrated Program, Department of the Environment, Ottawa.

A. R. Thompson, Faculty of Law, University of British Columbia, Vancouver.

R. Weir, Environmental Impact Assessment Division, Department of the Environment, Ottawa.

C. List of U.S. and Alaskan respondents

Washington, D.C. interviews

Reed Bohne, Office of Ocean Resources Coordination and Assessment, Office of Coastal Zone Management, National Oceanic and Atmospheric Administration, Department of Commerce.

William Brah, Office of Coastal Zone Management, National Oceanic and Atmospheric Administration, Department of Commerce.

Piet DeWitt, Bureau of Land Management, Department of the Interior.

Charles N. Ehler, Office of Ocean Resources Coordination and Assessment, Office of Coastal Zone Management, National Oceanic and Atmospheric Administration, Department of Commerce.

Peter Holmes, Natural Resources Defense Council Inc.

Carolita Kalluar, Program Coordinator, Alaska Outer Continental Shelf Office, Bureau of Land Management, Department of the Interior.

George Kinter, Office of Ocean Resources Coordination and Assessment, Office of Coastal Zone Management, National Oceanic and Atmospheric Administration, Department of Commerce.

Richard Legatski, National Ocean Industries Association.

Peter McAvoy, Office of Coastal Zone Management, National Oceanic and Atmospheric Administration, Department of Commerce.

Gerry Mylroie, Coastal Energy Impact Program, Office of Coastal Zone Management, National Oceanic and Atmospheric Administration, Department of Commerce.

Paul R. Sang, Office of the Secretary, Department of the Interior.

Michele M. Tetley, Outer Continental Shelf Office, Department of the Interior.

Alaska interviews

Bruce Baker, Division of Policy Development and Planning, Office of the Governor, State of Alaska, Juneau, Alaska.

David Benton, Friends of the Earth, Anchorage, Alaska.

Veronica Clark, Division of Community Planning, Department of Community and Regional Affairs, State of Alaska, Juneau, Alaska.

Margie Gibson, Friends of the Earth, Anchorage, Alaska.

Judy Gottlieb, Environmental Assessment Division, Alaska Outer Continental Shelf Office, Bureau of Land Management, Anchorage, Alaska.

Richard Logan, Habitat Protection Section, Department of Fish and Game, State of Alaska, Juneau, Alaska.

John Miller, Division of Minerals and Energy Management, Department of Natural Resources, State of Alaska, Anchorage, Alaska.

Robert Mintz, Trustees for Alaska, Anchorage, Alaska.

Brian F. School, Conservation Division, U.S. Geological Survey, Anchorage, Alaska.

Mark Stephens, Division of Community Planning, Department of Community and Regional Affairs, State of Alaska, Anchorage, Alaska.

Lance L. Trasky, Habitat Protection Section, Department of Fish and Game, State of Alaska, Anchorage, Alaska.

Murray Walsh, Office of Coastal Management, Office of the Governor, State of Alaska, Juneau, Alaska.

References

Alaska Highway Pipeline Panel. 1977. *Initial environmental evaluation of the proposed Alaska Highway gas pipeline, Yukon Territory*. Prepared by Interdisciplinary Systems Ltd. Winnipeg, Manitoba: Alaska Highway Pipeline Panel.

American Society of Planning Officials. 1977. *Source book and workshop materials*. An ASPO Training Project: Onshore Impacts of Outer Continental Shelf Oil and Gas Development. Chicago, Illinois: American Society of Planning Officials.

Baker, J.M. 1977. Response of salt marsh vegetation to oil spills and refinery effluents. In *Ecological processes in coastal environments.* The First European Ecological Symposium and the 19th Symposium of the British Ecological Society, Norwich, 12-16 September 1977. Oxford, England: Blackwell Scientific Publications.

———, and K. Hisock. 1980. *Predicting the impact of oil terminal development on the in-shore marine environment*. Pembroke, Wales: British Field Studies Council, Oil Pollution Research Unit, Orielton Field Centre.

Baldwin, P. L., and M.F. Baldwin. 1975. *Onshore planning for offshore oil: lessons from Scotland*. Washington, D.C.: The Conservation Foundation.

Balneaves, E. 1979. Fortunes of the Shetlands: impact of an industrial revolution. *The Geographical Magazine* 50: 723-729.

Batey, P.W.J., and M.J. Breheny, eds. 1978. Systematic methods in British planning practice. *Town Planning Review* 39: 257-318, 445-518.

Berger, T.R. 1977. *Northern frontier, northern homeland: the report of the Mackenzie Valley Pipeline Inquiry*. 2 vols. Ottawa, Ontario: Supply and Services Canada.

Berry, R.J., and J.L. Johnston. 1980. *The natural history of Shetland*. St James Place, London: William Collins and Sons.

Biderman, J.C., and W.H. Drury. 1980. *The effects of low levels of oil on aquatic birds: a non-technical summary of research activities, FY76 through FY78*. Washington, D.C.: U.S., Department of the Interior, United States Fish and Wildlife Service, Biological Services Program.

Boeri, D. 1980. Oil on troubled Alaskan waters. *New York Times Magazine* (9 November): 132-149.

British Field Studies Council, Oil Pollution Research Unit. n.d. *Annual report: 1977-1978*. Pembroke, Wales: British Field Studies Council, Oil Pollution Research Unit, Orielton Field Centre.

Brother, P.C., and R.D. Legge, eds. 1980. *Contact* 12(2). Special Issue on Environmental Issues and Human Values. Waterloo: University of Waterloo, Faculty of Environmental Studies.

Clark, B.D., K. Chapman, R. Bisset, and P. Wathern. 1976. *Assessment of major industrial applications: a manual*. Research Report No. 13. Prepared at the Department of Geography', University of Aberdeen. Aberdeen,Scotland: G.B., Department of the Environment.

Coull, J.R., J.H. Goodlad, and G.T. Sheves. 1979. *The fisheries in the Shetland area; A study in conservation and development*. Sponsored by the Shetland Islands Council and the White Fish Authority. Aberdeen, Scotland: University of Aberdeen, Department of Geography.

Crommelin, M. 1974. Offshore oil and gas rights: a comparative study. *Natural Resources Journal* 14: 457-500.

Dames, J., and S. Moore. 1978. *Beaufort Sea petroleum development scenarios for the state-federal and federal outer continental shelf: executive summary*. Socio-economic Studies Program, Technical Report No. 6A. Anchorage, Alaska: U.S., Department of the Interior, Bureau of Land Management, Alaska Outer Continental Shelf Office.

David M. Dornbusch and Company, Inc. 1976. *Management of OCS-related industrial development: a guide for Alaskan coastal communities*. A report prepared for the Division of Community Planning, Department of Community and Regional Affairs, State of Alaska.

Dean, F.E. 1979. Here? There? or Anywhere? The environmental problems associated with the onshore development of the northern basin gas fields. Presented to the annual general meeting of the Institute of Gas Engineers, Scottish Section, in Aviemore, Scotland.

———. n.d. *The onshore problems of offshore exploration and exploitation*. London, England: British Gas Corporation.

———, and G. Graham. 1979. Environmental impact analysis in the British gas industry. Transmitted by the Government of the United Kingdom to the Economic Commission for Europe's Seminar on Environmental Impact Assessment, 24-29 September 1979, Villaeh, Austria.

———, G. Graham, and J.A. Ward. 1979. *North Sea gas and the environment: value for money?* Cardiff,Wales: Institute of Gas Engineers, Wales Section.

Dicks, B. n.d. *The applicability of the Milford Haven experience for new oil terminals*. Pembroke,Wales: British Field Studies Council, Oil Pollution Research Unit, Orielton Field Centre.

Dudley, G. 1968. The problem of oil pollution in a major oil port. In *Biological effects of oil pollution on littoral communities*. Proceedings of the Conference on Biological Effects of Oil Pollution on Littoral Communities, February 1968, ed. J.D. Carthy and D.R. Arthur, pp. 21-29. (Supplement to *Field Studies*, Vol. 2.)

______. 1976. The incidence and treatment of oil pollution in oil ports. In *Marine ecology and oil pollution*, ed. J.M. Baker, pp. 27-40. Barking, England: Applied Science Publishers.

Duke, J.H., J.M. Bishop, D.J. Basta, and C.N. Ehler. 1980. *Development of a simple pollutant transport model for the Gulf of Mexico*. Washington, D.C.: U.S., Department of Commerce, National Oceanic and Atmospheric Administration, Office of Coastal Zone Management, Office of Ocean Resources Coordination and Assessment.

Dunnet, G.M. 1980. Personal communication.

Economist Intelligence Unit Ltd. 1975. *Buchan impact study*, Parts 1 and 2. London, England: The Economist Intelligence Unit Ltd.

Farlow, J.S., R. Wilder, and C. Potter, eds. 1980. *Oil spill debris: where to put the waste*. Interagency Energy/Environment Research and Development Program Report. Cincinnati,Ohio: U.S., Environmental Protection Agency, Office of Research and Development, Industrial Environmental Research Laboratory.

Fisher, D.W., and D. von Winterfeldt. 1978. Setting standards for chronic oil discharges in the North Sea. *Journal of Environmental Management* 7: 177-199.

Foxton, P. 1979. The SOTEAG Programme of monitoring in Sullom Voe. Presented to the SMBA/SOTEAG Scientific Symposium, April 1979, Oban, Scotland.

Freeman, M.M.R., A.C. Cooke, and F.H. Schwartz, eds. 1976. *Inuit land use and occupancy project*. 3 vols. Prepared by Milton Freeman Research Limited for the Department of Indian Affairs and Northern Development. Ottawa, Ontario: Supply and Services Canada.

Gammack, J., and M.G. Richardson. 1980. *A compendium of ecological information for oil spill contingency planning, Shetland*. Draft. Edinburgh, Scotland: Nature Conservancy Council.

Gavin, A. 1978. *Wildlife of the North Slope: a ten-year study 1969-1978*. Atlantic Richfield Company.

Goodier, R., and A.A. Whyte. 1976. *Compendium of ecological studies of the Scottish coast relevant to development planning*. Nature Conservancy Council.

Gowen, A.M., M.J. Goetz, and I.M. Waitsman. 1980. *Choosing offshore pipeline routes: problems and solutions*. Interagency Energy/Environment Research and Development Program Report. Cincinnati, Ohio: U.S., Environmental Protection Agency, Office of Research and Development, Industrial Environmental Research Laboratory.

G.B., Central Office of Information. 1979. *Environmental planning in Britain*. Reference Pamphlet 9. London, England: Her Majesty's Stationery Office.

G.B., Department of Energy. 1978. *A North Sea gas gathering system*. Energy Paper No. 44. London, England: Her Majesty's Stationery Office.

G.B., Department of the Environment. 1976. *Accidental oil pollution of the sea*. Pollution Paper No. 8. London, England: Her Majesty's Stationery Office.

______. 1977. Press release (28 July).

G.B., Scottish Information Office, Reference Unit. 1974. *Scottish Office brief on North Sea oil*. Edinburgh, Scotland: Scottish Information Office, Reference Unit.

G.B., Scottish Office. 1974. *North Sea oil and the environment*. Edinburgh, Scotland: Scottish Office.

G.B., Scottish Office, Scottish Development Department. 1974. *North Sea oil and gas: coastal planning guidelines*. Edinburgh, Scotland: Scottish Development Department.

______. 1975. *North Sea oil and gas developments in Scotland: a physical planning resume*. Edinburgh, Scotland: Scottish Development Department.

______. 1976. *North Sea oil and gas developments: environmental impact analysis, Scottish experience, 1973-1975*. Edinburgh, Scotland: Scottish Development Department.

______. 1977a. *Electricity*. National Planning Series, Land Use Summary Sheet No. 7. Edinburgh, Scotland: Scottish Development Department.

______. 1977b. *National planning guidelines for large industrial sites and rural conservation*. Edinburgh, Scotland: Scottish Development Department.

______. 1977c. *National Planning Guidelines for petrochemical developments*. Edinburgh, Scotland: Scottish Development Department.

______. 1977d. *Oil, gas and petrochemicals*. National Planning Series, Land Use Summary Sheet No. 6. Edinburgh, Scotland: Scottish Development Department.

______. 1977e. *Planning Information Note A3B*. Edinburgh, Scotland: Scottish Development Department.

______. 1977f. *Planning Information Note A10*. Edinburgh, Scotland: Scottish Development Department.

______. 1977g. *Planning Information Note A11 (a-m)*. Edinburgh, Scotland: Scottish Development Department.

G.B., Scottish Office, Scottish Development Department, Central Research Unit. n.d. *Social consequences of oil developments*. Summary of University of Aberdeen research report. Edinburgh, Scotland: Scottish Development Department.

G.B., Scottish Office, Scottish Economic Planning Department. 1979. *North Sea oil information sheet*. Edinburgh, Scotland: Scottish Economic Planning Department.

Greenpeace. 1980. *Oil drilling in the Beaufort Sea; Greenpeace policy statement and research report*. Anchorage, Alaska: Greenpeace.

Gronneberg, R., ed. 1978. *Island futures: Scottish devolution and Shetland's constitutional alternatives*. Sandwick, Shetland Islands: Thuleprint.

Holling, C.S. 1978. *Adaptive environment assessment and management*. Sponsored by the United Nations Environmental Program and developed at the Workshop on Adaptive Assessment of Ecological Policies, International Institute for Applied Systems Analysis, 1977, Laxenburg, Austria. New York: John Wiley and Sons.

Houston, L.H. 1977. Legal problems of land use and acquisition for oil-related development. In *Manual of United Kingdom oil and gas law*, ed. T. Baintith and J. Willoughby, ch. 4. London: Oyez Publications.

______. 1978. Planning legislation in Britain. Presented to the Petroleum Law Seminar, January 1978, Churchill College, Cambridge, England. Organized by the International Bar Association, Section on Business Law, Committee on Energy and Natural Resources.

Hunt, C.D. 1978. Approaches to native land settlements and implications for northern land use and resource management policies. In *Northern Transitions*, Vol. 2: Second National Workshop on People, Resources and the Environment North of 60°, ed. R.F. Keith and J.B. Wright, pp. 5-41. Ottawa, Ontario: Canadian Arctic Resources Committee.

______, and A.R. Lucas. 1980. *Environmental regulation; its impact on major oil and gas projects — oil sands and arctic*. Calgary, Alberta: University of Calgary, Canadian Institute of Resources Law.

International Conference on Oil and the Environment. 1980. Abstracts of papers presented at the conference, Onshore Impact of Offshore Oil, 29 September-1 October 1980, University of Edinburgh. Edinburgh, Scotland: International Conference on Oil and the Environment. Titles include: Environmental aspects of the development of the North Sea oil; The North Sea and its coastal lands: an environmental overview; Control strategies: conservation; Europe and environmental assessment; St Fergus: visual and design considerations; The impact of North Sea oil on Orkney; The Shetland Islands and North Sea oil.

James Lindsay and Associates. 1978. *Beaufort Sea petroleum development scenario: summary of socio-economic impacts*. Socio-economic Studies Program, Technical Report No. 23. Anchorage, Alaska: U.S., Department of the Interior, Bureau of Land Management, Alaska Outer Continental Shelf Office.

Keith, R.F., and J.B. Wright, eds. 1978. *Northern Transitions*, Vol. 2: Second National Workshop on People, Resources and the Environment North of 60°. Ottawa, Ontario: Canadian Arctic Resources Committee.

Kellas, J.G. 1980. *Modern Scotland*. 2d ed. London, England: George Allen & Unwin.

Longley, W. L., R. Snyder, and B. Snyder. 1978. *Managing oil and gas activities in coastal environments*. Washington, D.C.: U.S., Department of the Interior, United States Fish and Wildlife Service, Biological Services Program.

MacPherson, G.S., and J. Bernstein. 1980. *Outer continental shelf oil and gas activities in the Pacific (southern California) and their onshore impacts: a summary report, May 1980*. Open-File Report 80-645. Washington, D.C.: U.S., Department of the Interior, U.S. Geological Survey.

Marshall, B., J.D. Okill, and I. Sandison. 1980. *Shetland bird report 1979*. Lerwick, Shetland Islands: Shetlands Bird Club.

Milford Haven Standing Conference on Anti-Oil Pollution. 1977. *Port of Milford Haven anti-oil pollution plan*. Milford Haven Standing Conference on Anti-Oil pollution.

Milner, C. 1979. Shetland ecology surveyed: natural history in the balance. *The Geographical Magazine* 50(11): 730-736.

MITRE Corporation. 1979a. *Outer Continental Shelf Oil and Gas Information Program: Alaska index (December 1974-June 1979); An annotated guide to reports and documents used by the federal government in the decision-making process for leasing, exploration, development, and production on the outer continental shelf*. Open-File Report 79-1345. Washington, D.C.: U.S., Department of the Interior, U.S. Geological Survey.

———. 1979b. *Outer Continental Shelf Oil and Gas Information Program: directory to federal, state and local OCS-related activities and contacts.* Open-File Report 79-1481. Prepared for the U.S. Geological Survey in co-operation with the Council on Environmental Quality. Washington, D.C.: U.S., Department of the Interior, U.S. Geological Survey.

Morehouse, T.A., and L.C. Leask. 1978. *Governance in the Beaufort Sea region: petroleum development and the North Slope Borough.* Socio-economic Studies Program, Technical Report No. 16. Anchorage, Alaska: U.S., Department of the Interior, Bureau of Land Management, Alaska Outer Continental Shelf Office.

National Research Council, Commission on Natural Resources, Environmental Studies Board, Committee to Evaluate Outer Continental Shelf Environmental Studies. 1978. *OCS oil and gas: an assessment of the Department of the Interior environmental studies program.* A report to the Bureau of Land Management, Department of the Interior. Washington, D.C.: National Academy of Sciences.

Natural Environment Research Council, Experimental Cartography Unit. 1978. *Land use mapping by local authorities.* London, England: Architectural Press.

Nature Conservancy Council. 1975. *Nature Conservancy Council: first report.* London, England: Her Majesty's Stationery Office.

Neakok, B. 1977. Memorandum statement of the environmental protection policy of the North Slope Borough. Memorandum from B. Neakok, Director of Conservation and Environmental Security, to Mayor Eben Hopson. Barrow, Alaska: North Slope Borough.

Nelson, J.G. 1975. *Arctic renewable resources: summary and recommendations.* Renewable Resources Project, Vol. 11. Ottawa, Ontario: Inuit Tapirisat of Canada.

———. 1980. Northern pipelines and environmental strategies: an environmental management model. Presented to the Alberta Society of Environmental Biologists, Edmonton, Alberta.

———, J.D. Day, and S. Jessen. 1980. *Assessing the environmental protection regulatory system for the Nanticoke industrial complex.* Technical Report No. 1. Ottawa, Ontario: Economic Council of Canada.

New England River Basins Commission. 1979a. *Annual report.* Boston, Massachusetts: New England River Basins Commission.

———. 1979b. *Strategies for state participation in OCS exploration decisions.* Proceedings of the 22 March 1979 conference sponsored by the Coastal Programs Division, New England River Basins Commission, and Resource and Land Investigations, U.S. Geological Survey. Boston, Massachusetts: New England River Basins Commission.

———. 1980a. *State participation in OCS development and production decisions.* Boston, Massachusetts: New England River Basins Commission.

———. 1980b. *Strategies for state participation in OCS exploration decisions.* Boston, Massachusetts: New England River Basins Commission.

Noble, J.H., J.S. Banta, and J.S. Rosenberg, eds. 1977. *Groping through the maze: foreign experience applied to the U.S. problem of coordinating development controls.* A Conservation Foundation report from the International Comparative Land-Use Program. Washington, D.C.: The Conservation Foundation.

North Slope Borough. 1979. *Coastal Management Program, Mid-Beaufort Segment.* Draft. Barrow, Alaska: North Slope Borough.

———. 1980. NSB withdraws coastal zone plan: industry opposition prevails. *The Arctic Coastal Zone Management Newsletter* 26:1-3.

Norton, D.W., and G. Weller. 1980. Impact of a $16M budget level for FY81 on arctic biological and physical programs of OCSEAP. Internal memorandum to H.N. Bruce, Director, Outer Continental Shelf Environmental Assessment Program.

O'Riordan, T. 1979. The role of environmental quality objectives in the politics of pollution control. In T. O'Riordan and R.C. D'Arge. 1979. *Progress in resource management and environmental planning*, Vol. 1, pp. 221-258. New York: John Wiley and Sons.

Peterson, E.B., and J.B. Wright, eds. 1978. *Northern Transitions*, Vol. 1: Northern Resource and Land Use Policy Study. Ottawa, Ontario: Canadian Arctic Resources Committee.

Rees, W.E. 1978. Development and planning north of 60°: past and future. In *Northern Transitions*, Vol. 2: Second National Workshop on People, Resources and the Environment North of 60°, ed. R.F. Keith and J.B. Wright, pp. 42-62. Ottawa, Ontario: Canadian Arctic Resources Committee.

Rivkin, M.D. 1977. *Negotiated development: a breakthrough in environmental controversies.* Washington, D.C.: The Conservation Foundation.

Sadler, P.G. 1977. The economic, social and political impacts of large-scale industrial development of rural areas. In *The future of upland Britain.* CAS Paper No. 2. Reading, England.

Schneider, D.M., ed. 1977. *Planning for onshore development: discussion paper.* An ASPO Training Project: Onshore Impacts of Outer Continental Shelf Oil and Gas Development. Chicago, Illinois: American Society of Planning Officials.

Shanks, L.R. 1978. *Coastal systems and management options related to outer continental shelf (OCS) development*. Washington, D.C.: U.S., Department of the Interior, United States Fish and Wildlife Service, Biological Services Program.

Shetland Islands Council. 1976. *Report of survey*, Vol. 1: Summary, Shetland Structure Plan. Lerwick, Shetland Islands: Shetland Islands Council.

———. n.d. *Shetland Islands anti-oil pollution scheme*. Lerwick, Shetland Islands: Shetland Islands Council.

Shetland Islands Council, Planning Department. 1978. *Written statement of policies and proposals, Shetland Structure Plan*, Vol. 4. Lerwick, Shetland Islands: Shetland Islands Council, Planning Department.

Shetland Islands Council, Research and Development Department. 1980. *Shetland in statistics*, No. 9. Lerwick, Shetland Islands: Shetland Islands Council, Research and Development Department.

State of Alaska, Alaska Coastal Policy Council. 1980. Agenda: Alaska Coastal Policy Council meeting, 16-17 September 1980. Item F: Compensation for public participation. Juneau, Alaska: State of Alaska, Office of the Governor, Division of Policy Development and Planning.

State of Alaska, Department of Commerce and Economic Development, Division of Economic Enterprise, and State of Alaska, Department of Environmental Conservation. 1979. *Directory of permits*. Juneau, Alaska: State of Alaska, Department of Commerce and Economic Development, Division of Economic Enterprise.

State of Alaska, Department of Community and Regional Affairs, Division of Community Planning. 1977. *Planning for offshore oil development: economic forecasts, Lower Cook Inlet and lease sale*. Juneau, Alaska: State of Alaska, Department of Community and Regional Affairs, Division of Community Planning.

———. 1978. *Planning for offshore oil development: Gulf of Alaska OCS handbook*. Juneau, Alaska: State of Alaska, Department of Community and Regional Affairs, Division of Community Planning.

———. 1979. *Alaska Coastal Management Program: district guidebook*, Vols. 1-5 and Appendix. Anchorage, Alaska: State of Alaska, Department of Community and Regional Affairs.

State of Alaska, Department of Natural Resources. 1980. *State Five-year Leasing Program presented to the second session eleventh Alaska legislature*. Juneau, Alaska: State of Alaska, Department of Natural Resources.

State of Alaska, Department of Natural Resources, Office of the Commissioner. 1980. *State Five-year Oil and Gas Leasing Schedule revised.* Press release (18 July). Juneau, Alaska: State of Alaska, Department of Natural Resources.

State of Alaska, Office of the Governor. 1979. *Governor's Administrative Order No. 55: State Major Project Review Process.* Juneau, Alaska: State of Alaska, Office of the Governor.

State of Alaska, Office of the Governor, Division of Policy Development and Planning, Office of Coastal Management. 1980a. *Annual report 1979.* Juneau, Alaska: State of Alaska, Office of the Governor, Division of Policy Development and Planning, Office of Coastal Management.

———. 1980b. Consistency with Alaska Coastal Management Program. Unpublished paper presented to the 28-29 October meeting of the Alaska Coastal Policy Council.

———. 1980c. *Special areas in the Alaska coastal zone: abstracts of proposals.* Juneau, Alaska: State of Alaska, Office of the Governor.

———, and U.S., Department of Commerce, National Oceanic and Atmospheric Administration, Office of Coastal Zone Management. 1979. *State of Alaska Coastal Management Program and final environmental impact statement.* Washington, D.C.: U.S., Department of Commerce.

State of Alaska, Office of the Governor, Governor's Agency Advisory Committee on Leasing. 1980. *A social, economic and environmental analysis of a state oil and gas lease sale in Upper Cook Inlet.* Preliminary draft. Juneau, Alaska: State of Alaska, Office of the Governor.

Sullom Voe Environmental Advisory Group. 1976. *Oil terminal at Sullom Voe: environmental impact assessment.* Sandwick, Shetland Islands: Sullom Voe Environmental Advisory Group.

Sullom Voe Harbour Authority. 1980. *Port information and tidal predictions for the year 1980.* Seallaness, Graven, Shetland Islands: Sullom Voe Harbour Authority.

Sullom Voe Oil Spill Advisory Committee. n.d. *Sullom Voe harbour oil spill plan.* Sullom Voe Harbour Authority, and the Operator of the Sullom Voe Oil Terminal.

Terrell, T.T. 1979. *Physical regionalization of coastal ecosystems of the United States and its territories.* Washington, D.C.: U.S., Department of the Interior, United States Fish and Wildlife Service, Biological Services Program.

Trimble, N.F. 1977. *Planning for offshore oil development: the use of mathematical models to evaluate OCS impact on Alaska*. Juneau, Alaska: State of Alaska, Department of Community and Regional Affairs, Division of Community Planning.

Trustees for Alaska. 1979. *Newsletter* (May). Anchorage, Alaska: Trustees for Alaska.

U.S., Council on Environmental Quality. 1974. *OCS oil and gas: an environmental assessment; A report to the president*. Washington, D.C.: Government Printing Office.

______, and U.S., Department of Commerce, National Oceanic and Atmospheric Administration, Office of Coastal Zone Management, Office of Ocean Resources Coordination and Assessment. n.d. *Eastern United States coastal and ocean zone resource use mapping project*. Washington, D.C.: U.S., Department of Commerce, National Oceanic and Atmospheric Administration, Office of Coastal Zone Management, Office of Ocean Resources Coordination and Assessment.

U.S., Department of Commerce, National Oceanic and Atmospheric Administration. 1976. Coastal Energy Impact Program: proposed regulations for financial assistance to coastal states. *Federal Register* 41 (206): 46723-46740.

______. 1980. *The NOAA story*. Washington, D.C.: U.S., Department of Commerce, National Oceanic and Atmospheric Administration.

U.S., Department of Commerce, National Oceanic and Atmospheric Administration, Arctic Project Office. 1979. *Environmental stipulations relating to OCS development of the Beaufort Sea: proceedings of a synthesis meeting of OCSEAP investigators*. Special Bulletin No. 25. Fairbanks, Alaska: Arctic Project Office, University of Alaska.

U.S., Department of Commerce, National Oceanic and Atmospheric Administration, Interagency Committee on Ocean Pollution Research, Development and Monitoring. 1979. *Catalog of federal ocean pollution research, development and monitoring programs, fiscal years 1978-80*. Working Paper 1 for the Federal Plan for Ocean Pollution Research, Development and Monitoring, Fiscal Years 1979-83. Washington, D.C.: U.S., Department of Commerce, National Oceanic and Atmospheric Administration.

U.S., Department of Commerce, National Oceanic and Atmospheric Administration, Office of Coastal Zone Management. n.d. *Draft environmental impact statement: rules and regulations for implementing the Coastal Energy Impact Program, section 308, the Coastal Zone Management Act of 1972 as amended*. Washington, D.C.: U.S., Department of Commerce, National Oceanic and Atmospheric Administration, Office of Coastal Zone Management.

U.S., Department of Commerce, National Oceanic and Atmospheric Administration, Office of Coastal Zone Management, Office of Ocean Resources Coordination and Assessment. n.d. *Mission statement: Office of Ocean Resources Coordination and Assessment*. Washington, D.C.: U.S., Department of Commerce, National Oceanic and Atmospheric Administration.

U.S., Department of the Interior. 1980. *Secretarial issue document: 5-year OCS Oil and Gas Leasing Program*. Washington, D.C.: U.S., Department of the Interior.

U.S., Department of the Interior, Bureau of Land Management. 1978. *Study design for resource management decisions: OCS oil and gas development and the environment*. Washington, D.C.: U.S., Department of the Interior, Bureau of Land Management.

______. 1979a. *Final environmental impact statement: proposed federal state oil and gas lease sale Beaufort Sea*, Vols. 1-3. Washington, D.C.: Government Printing Office.

______. 1979b. *Intergovernmental Planning Program for OCS Oil and Gas Leasing, Transportation and Related Facilities*. Washington, D.C.: Government Printing Office.

______. 1980a. *Draft supplement to the final environmental impact statement*. Washington, D.C.: U.S., Department of the Interior, Bureau of Land Management.

______. 1980b. *Final environmental statement: proposed Five-year OCS Oil and Gas Lease Sale Schedule March 1980 - February 1985*. Washington, D.C.: U.S., Department of the Interior, Bureau of Land Management.

______. 1980c. *Prospectus of BLM marine research 1980-1985*. Washington, D.C.: U.S., Department of the Interior, Bureau of Land Management.

U.S., Department of the Interior, Bureau of Land Management, Alaska Outer Continental Shelf Office. 1977. *Baseline studies of the physical and man-made environment: the Beaufort Sea region*. Socio-economic Studies Program, Technical Report No. 5. Anchorage, Alaska: U.S., Department of the Interior, Bureau of Land Management, Alaska Outer Continental Shelf Office.

______. n.d. *Socio-economic impacts of selected foreign OCS developments*. Socio-economic Studies Program, Technical Report No. 28. Anchorage, Alaska: U.S., Department of the Interior, Bureau of Land Management, Alaska Outer Continental Shelf Office.

U.S., Department of the Interior, Bureau of Land Management, and State of Alaska, Department of Natural Resources. 1979. *Federal/State Joint Beaufort Sea Oil and Gas Lease Sale*.

U.S., Department of the Interior, Office of the Secretary. 1980a. *Five-year OCS Leasing Schedule approved*. News release (18 June). Washington, D.C.: U.S., Department of the Interior.

———. 1980b. *OCS Five-year Leasing Program decision announced.* News release (8 March). Washington, D.C.: U.S., Department of the Interior.

U.S., Department of the Interior, U.S. Geological Survey, Conservation Division. 1980. *Gulf of Alaska outer continental shelf orders government oil and gas lease operations*. Reston, Virginia: U.S., Department of the Interior, U.S. Geological Service, Conservation Division.

U.S., Department of the Interior, U.S. Geological Survey, Office of Deputy Division Chief for Offshore Minerals Regulation. 1980. *The use of best available and safest technologies (BAST) during oil and gas drilling and producing operations of the outer continental shelf (OCS)*. Reston, Virginia: U.S., Department of the Interior, U.S. Geological Survey, Office of Deputy Division Chief for Offshore Minerals Regulation.

U.S., General Accounting Office. 1979. *General Accounting Office report to the secretaries of the Interior and Commerce: benefits derived from the Outer Continental Shelf Environmental Studies Program are questionable*. Washington, D.C.: U.S., General Accounting Office.

U.S., House of Representatives, Select Committee on the Outer Continental Shelf. 1980. *Final report on the activities of the select committee on the Outer Continental Shelf including proposed legislative changes, 96th Congress*. Washington, D.C.: Government Printing Office.

University of Aberdeen, Department of Geography. 1971. *Ecological survey; Moray Firth development*. Interim Report No. 3. Aberdeen, Scotland: University of Aberdeen, Department of Geography.

———. 1980. *Ecological survey; Moray Firth development*. Interim Report No. 10. Aberdeen, Scotland: University of Aberdeen, Department of Geography.

University of Aberdeen, Department of Political Economy. 1978. *The economic impact of North Sea oil on Scotland: final report to the Scottish Economic Planning Department on a study conducted within the Department of Political Economy, University of Aberdeen, 1973-77*. Edinburgh, Scotland: Her Majesty's Stationery Office.

University of Alaska, Arctic Environmental Information and Data Center. 1979. *Bibliography of products, Alaska Coastal Management Program*. Juneau, Alaska: State of Alaska, Office of the Governor, Division of Policy Development and Planning, Office of Coastal Management.

University of Edinburgh, Tourism and Recreation Research Unit. 1976. *Research study into provision for recreation and islands*. Research Report No. 22. Edinburgh, Scotland: University of Edinburgh.

Usher, P.J. 1970. *The Bankslanders: economy and ecology of a frontier trapping community*, Vol. 2. Prepared for the Northern Science Research Group, Department of Indian Affairs and Northern Development. Ottawa, Ontario: Information Canada.

Wheeler, R.B. 1978. *The fate of petroleum in the marine environment*. Special Report. Exxon Production Research Company.

Williams, D.C., and K.B. Williams. 1979. *Onshore impacts of offshore oil: a user's guide to assessment methods*. Washington, D.C.: Government Printing Office.

Young, E., and P. Fricke, eds. 1975. *Sea use planning*. Fabian Tract 437. London, England: Fabian Society.

Index

A. Legislation

United States

United Kingdom

B. Government organizations

United States

Federal

Alaska

United Kingdom

C. Non-government organizations

United States

United Kingdom

United Kingdom (Cont'd)

D. Programmes